# Workbook to Accompany
# The Firefighter's Handbook:

## *Essentials of Firefighting and Emergency Response*

### Second Edition

# Workbook to Accompany The Firefighter's Handbook:

## *Essentials of Firefighting and Emergency Response*

## Second Edition

**Andrea A. Walter**

**Marty L. Rutledge**

**Dennis Childress**

**Dave Dodson**

**Chris Hawley**

**DELMAR**

**THOMSON LEARNING** ™

Australia Canada Mexico Singapore Spain United Kingdom United States

**THOMSON**

**DELMAR LEARNING**

Workbook to Accompany The Firefighter's Handbook: Essentials of Emergency Response, Second Edition

Andrea A. Walter, Marty L. Rutledge, Dennis Childress, Dave Dodson, and Chris Hawley

**Vice President, Technology and Trades SBU:**
Alar Elken

**Editorial Director:**
Sandy Clark

**Acquisitions Editor:**
Alison S. Weintraub

**Developmental Editor:**
Jennifer A. Thompson

**Marketing Director:**
Cyndi Eichelman

**Channel Manager:**
Bill Lawrensen

**Marketing Coordinator:**
Mark Pierro

**Production Director:**
Mary Ellen Black

**Production Editor:**
Barbara L. Diaz

**Art/Design Specialist:**
Rachel Baker

**Technology Project Manager:**
Kevin Smith

**Editorial Assistant:**
Stacey Wiktorek

Library of Congress Cataloging-in-Publication Data:

Firefighter's handbook : essentials of firefighting and emergency response.—2nd ed.
        p. cm.
        Includes bibliographical references and index.
        ISBN 1-4018-3575-9 (alk. paper)
        1. Fire extinction—Handbooks, manuals, etc.
I. Delmar Publishers.
    TH9151.F458 2005
    628.9′25—dc22
                        2003066273
ISBN: 1-4018-3578-3

## NOTICE TO THE READER

Publisher does not warrant or guarantee any of the products described herein or perform any independent analysis in connection with any of the product information contained herein. Publisher does not assume, and expressly disclaims, any obligation to obtain and include information other than that provided to it by the manufacturer.

The reader is expressly warned to consider and adopt all safety precautions that might be indicated by the activities herein and to avoid all potential hazards. By following the instructions contained herein, the reader willingly assumes all risks in connection with such instructions.

The publisher makes no representation or warranties of any kind, including but not limited to, the warranties of fitness for particular purpose or merchantability, nor are any such representations implied with respect to the material set forth herein, and the publisher takes no responsibility with respect to such material. The publisher shall not be liable for any special, consequential, or exemplary damages resulting, in whole or part, from the reader's use of, or reliance upon, this material.

# CONTENTS

# INTRODUCTION

This *Workbook* is designed to accompany *The Firefighter's Handbook: Essentials of Firefighting and Emergency Response, Second Edition,* as well as *The Firefighter's Handbook: Basic Essentials of Firefighting.* It is intended to be a companion text that will assist the reader in studying and reinforcing the information gained from either of these texts.

Readers and students will find that the *Workbook* questions come directly from the information and material presented in *The Firefighter's Handbook* and that the questions help reinforce key points and important information found within the body of the text. These study questions can be helpful to readers and students as a self-test to improve reading comprehension or as a review instrument for quizzes and tests.

The *Workbook* contains a variety of types of questions, including matching, fill-in-the-blank, multiple choice, and true or false. When completing the questions in the *Workbook,* readers and students should first try to complete the section without the assistance of *The Firefighter's Handbook,* then check the answers to all the questions provided at the end of the *Workbook.* This will help readers and students to get the most from the exercises found in the *Workbook.* The companion text should be used to look up answers that are incorrect or for questions the reader or student is unable to answer on the first try.

The *Workbook* draws questions from both Firefighter I and Firefighter II levels of information. In order to differentiate between these two levels, a special symbol **FFII** marks those questions pertaining to the Firefighter II level.

When used in concert with *The Firefighter's Handbook,* the *Workbook* will serve as a valuable learning tool for readers and students interested in improving their knowledge of firefighting and emergency response.

# OVERVIEW OF THE HISTORY, TRADITION, AND DEVELOPMENT OF THE AMERICAN FIRE SERVICE

*Photo courtesy of Marysville Fire Department (California)*

# QUESTIONS

## Matching

Match the correct term with the definitions provided.

    a. mission statement

    b. firemark

    c. triple combination engine company

    d. conflagration

    e. fire wardens

1. The _____ was placed on an occupancy to tell firefighters which insurance company held the insurance policy on the building.

2. _____ were contracted to walk streets in an effort to spot fire and report it.

3. A unit designed to carry water, pump it, and also carry hose is called a _____ .

4. A written declaration called a _____ describes the things a fire agency intends to do in protecting its citizenry.

5. A very large and destructive fire is termed a _____ and usually requires a great number of units to control.

## True or False

1. It is not important that all firefighters at all levels of the organization understand the motivation behind the department's mission statement.
   a. true    b. false

2. Mankind invented fire in the early years of prehistoric life.
   a. true    b. false

3. When insurance was first utilized in the 1600s, it was far from successful.
   a. true    b. false

4. The internal combustion engine became the power source of choice for fire apparatus just after World War II.
   a. true    b. false

5. Over the decades fire has been a weapon of war as well as a force of nature.
   a. true    b. false

## Multiple Choice

1. The first fire pump was based upon the simple principle of:
   a. centrifugal force
   b. the syringe-like plunger
   c. dynamic acceleration
   d. steam-driven pressure

2. The Maltese Cross is symbolic for the fire service and was developed by an organization called the:
   a. Roman Guard
   b. Red Cross
   c. Knights of Malta
   d. Maltese Falcons

3. In the 1700s groups of people began banding together to deal with their fire problems. These groups were the forerunners of which of the following groups?
   a. Roman Brigades
   b. Veterans of War
   c. Fire Wardens
   d. Volunteer Firefighters

4. Which of the following people is considered by many to be America's first fire chief?
   a. Ben Franklin
   b. Thomas Jefferson
   c. George Washington
   d. Lloyd Layman

5. Volunteer fire companies were becoming quite prolific around what time in the nation's history?
   a. 1600
   b. 1700
   c. 1800
   d. 1900

6. The first fire helmet was created by Andrew Gratacap and was made of which of the following materials?
   a. polycarbonate
   b. leather
   c. plastic
   d. fiberglass

7. In Chicago and Detroit around 1850, the volunteer fire companies became heavily involved in politics and physical competition among themselves. These times soon became known as which of the following eras?
   a. Volunteer Watchmen
   b. Fire Mark
   c. Rowdies and Rum
   d. Knights of Malta

8. A riot by firefighters in Ohio in the 1850s led to what new concept of firefighting?
   a. paid firefighters
   b. firefighter unions
   c. volunteer labor forces
   d. emergency medical services

9. The Dalmatian was first introduced into the fire service for what reason?

    a. to be a companion to twenty-four-hour firefighters

    b. to be a companion for the horses

    c. to aid in search and rescue

    d. to aid in public education

10. World War II brought many new innovations into the fire service. Which of the following is *not* one of those innovations from that time period?

    a. radios

    b. diesel engines

    c. synthetic hose

    d. better command structure

# FIRE DEPARTMENT ORGANIZATION, COMMAND, AND CONTROL

# QUESTIONS

## Matching

Match the correct term with the definitions provided.

a. NFPA 1001
b. engine company
c. staging area
d. task force
e. standard operating procedure
f. unity of command
g. strike team
h. company
i. span of control
j. operational periods
k. transfer of command
l. mutual aid
m. incident management system
n. group
o. division
p. branch

1. The standard that defines minimum qualifications for a firefighter and outlines the knowledge and skills necessary to be successful is _____.

2. A(n) _____ provides specific information and instruction on how a task or assignment is to be accomplished.

3. Having one designated leader in command of an operation is known as _____ and is meant to keep responders organized and working toward a common goal.

4. _____ is the ability of one individual to supervise a number of other people or units.

5. The process of passing command to another qualified person on the fireground is called _____ and must be done by giving the oncoming leader a situation or status report.

6. A(n) _____ is an organized, systematic method for the command, control, and management of an emergency incident.

7. A(n) _____ is a functional designation to conduct a specific task, such as search and rescue or ventilation.

8. A(n) _____ is any combination of single resources assembled for an assignment, such as responding to a report of a structure fire.

9. A(n) _____ is a set number of resources of the same type and kind, such as five engine companies or five wildland firefighting trucks.

10. In the incident management system, the _____ is responsible for all operations within an assigned geographic area.

11. Communities and jurisdictions share resources by creating reciprocal arrangements called _____ agreements.

12. At a longer term incident, the time frames for operations to be conducted by a certain team or apparatus are referred to as _____.

13. The _____ is the unit designation of a group of firefighters assigned to a piece of apparatus designed to deliver and apply water at the fire scene.

14. The _____ is the part of the operations section where apparatus and personnel assigned to the incident are available for deployment.

15. A(n) _____ is a team of firefighters with apparatus assigned to perform a specific function in a designated response area.

16. A(n) _____ is the command designation established to maintain span of control over a number of divisions, sectors, or groups.

17. Figure 2-1 shows the organizational structure for managing a large incident. The structure contains multiple levels of management. Match the organizational structure to the managerial levels.

    a. supervisors

    b. command

    c. leaders

    d. section chief

    e. directors

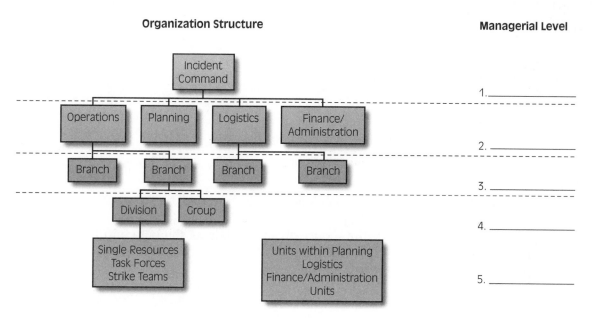

**Figure 2-1** Organizational structure for managing large incidents.

# True or False

1. The fire department mission statement should be communicated to the public.
    a. true    b. false

2. To accomplish its tasks, the fire department must have some type of organizational structure.
    a. true    b. false

3. One of the essential missions of all fire departments is the prevention of fires.
    a. true    b. false

4. Public fire/life safety education is not usually assigned to fire prevention since the primary function of prevention is code work.
    a. true    b. false

5. The Plans Division in a fire department is usually responsible for maintaining training records and activities.

    a. true     b. false

◫ 6. A mutual aid agreement is a reciprocal agreement.

    a. true     b. false

◫ 7. The most basic tactical unit on the fire scene is the strike team.

    a. true     b. false

8. The incident commander (IC) has the main responsibility for the safe control of an incident.

    a. true     b. false

9. NFPA 1901 states that pumpers can carry less than 300 gallons of water.

    a. true     b. false

10. When staging is used on a large incident, it is usually a function of the operations chief.

    a. true     b. false

## Multiple Choice

1. When viewing the organizational structure of a fire department, it is clear that its purpose is to accomplish all but which of the following?

    a. the assignment of responsibility for response areas

    b. to eliminate duplication of work and confusion

    c. to assign responsibility for specific activities

    d. the sectoring of staging areas on the fire scene

2. It is said that the key position of the fire service organization in its ability to fulfill its mission is which of the following?

    a. firefighter

    b. citizen

    c. governor

    d. taxpayer

3. According to the text, to effectively perform the duties of a successful firefighter at an emergency one must possess which of the following?

    a. strength and stamina

    b. knowledge and skills

    c. compassion and bravery

    d. speed and agility

4. What rank is most usually associated with the responsibility of all activities of the fire company?

    a. firefighter

    b. engineer

    c. captain

    d. battalion chief

5. A number of NFPA standards are used in the structure of the fire service organization. Which of the following standards best defines that of the company officer?

   a. 1001

   b. 1002

   c. 1010

   d. 1021

6. The typical fire prevention office is divided into two functions. Those are: (choose two)

   a. code enforcement

   b. training

   c. operations

   d. fire safety education

7. An organization outside the fire service has a great deal of responsibility in maintaining safety standards and the enforcement of those standards. Which of the following choices is that organization?

   a. public education

   b. OSHA

   c. city council

   d. educational system

8. Which two of the following choices would best describe where one would look for the procedures that address emergency operational tactics as designed by the fire department?

   a. rule books

   b. SOPs

   c. regulations

   d. bylaws

9. Which of the following outside agencies would assist the fire department in scene security?

   a. public works

   b. utility company

   c. police

   d. environmental protection

10. Which of the following outside agencies would assist the fire department in maintaining the public water system from which to draw water for firefighting?

   a. public works

   b. utility company

   c. police

   d. environmental protection

11. If at a hazardous materials fire the runoff was heading into the nearby creek, which of the following agencies would be contacted?

   a. social services

   b. utility company

   c. police

   d. environmental protection

⊞ 12. Many communities have agreements for assistance from each other in times of emergency. Which of the following terms best describes this type of agreement?

    a. assistance for hire

    b. mutual aid

    c. callback

    d. force hire

⊞ 13. Span of control is a management tool used by most fire departments, and it sets a given number of people that can work for one manager most effectively. What is the optimum number the text suggests for this management tool?

    a. two

    b. five

    c. seven

    d. ten

⊞ 14. Transfer of command is the passing on of vital information on the fire scene. How should this passing of information be best communicated?

    a. face to face

    b. radio

    c. paper transfer

    d. liaison

15. According to the text, to function properly the incident management system an agency uses should contain all but which of the following components?

    a. common terminology

    b. modular organization

    c. dedicated frequencies

    d. span of control

⊞ 16. Which of the following positions is not one commonly associated with command staff in the IMS?

    a. safety officer

    b. liaison

    c. public information officer

    d. planning officer

⊞ 17. On a large incident, who is responsible for the development of strategic goals?

    a. operations chief

    b. incident commander

    c. engine captain

    d. strike team leader

18. On larger incidents a staging area may be utilized. When assigned to staging and awaiting an assignment, what is the standard deployment time given any unit?

    a. three minutes

    b. five minutes

    c. ten minutes

    d. not specified

19. On an incident involving multiple communities and jurisdictions, what type of command is preferred?

    a. basic command

    b. sectional command

    c. unified command

    d. joint command

20. What is the acronym for the not-for-profit organization that uses a consensus process to develop model fire prevention codes and firefighting training and safety standards?

    a. OSHA

    b. NIOSH

    c. FDOC

    d. NFPA

21. The organizational structure in Figure 2-2 represents a vital role in the fire service. Of the following, which answer best fits the description of this structure?

    a. organizational structure for a greater alarm fire

    b. organizational structure for a small to medium department

    c. organizational structure for a medium to large department

    d. organizational structure for a multiagency response

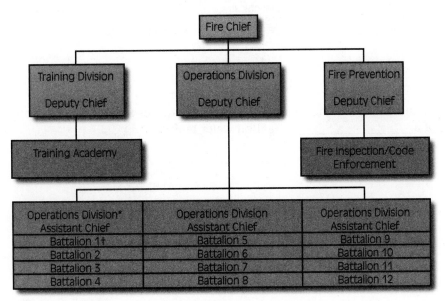

*Each division has one heavy rescue company.
† Battalion consists of four engine companies and one ladder company.

**Figure 2-2** Organizational structure.

22. Identify the chart in Figure 2-3.

    a. Incident Management System format

    b. command staff structure

    c. organizational chart for department structure

    d. staffing chart for department daily staffing

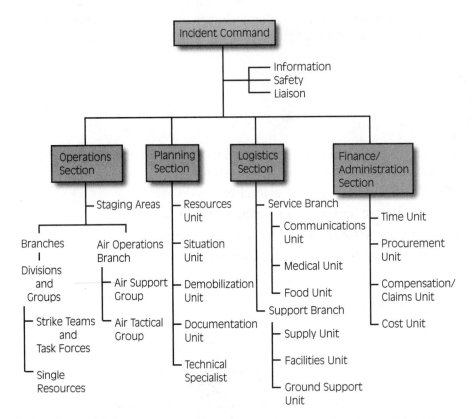

**Figure 2-3** Identify this chart.

23. NFPA 1901, Standard for Automotive Fire Apparatus, requires mobile water supply apparatus to have a minimum of a _____ gallon water tank.

    a. 300

    b. 500

    c. 800

    d. 1000

# Fill-in-the-Blanks

   a. command staff
   b. engine company
   c. finance section
   d. planning section
   e. truck company
   f. support
   g. rescue company
   h. operations section
   i. service
   j. incident commander
   k. mutual aid or assistance agreements

Ⓕⓘⓛ Organizing resources on the fireground can be a complex task if one is not familiar with the concepts and structure of the incident management system (IMS). The incident begins with the (1) _____ , which is designed to deliver firefighters, water, hose, and an attack to the fire upon arrival. Then the (2) _____ will arrive bringing forcible entry, ventilation, and ladders to the scene. If medical aid is needed on the scene, then a squad or (3) _____ will be called. As the incident escalates or grows, more resources will be ordered by the (4) _____ . While the incident commander is in charge of the overall incident, the (5) _____ will be responsible for the tactical assignments and their delegation to the field units. If it becomes a really large incident like some of the wildland conflagrations in the West, then command will need to set up a full staff. One of the general staff positions assigned to assist with the resource and statistical work will be the (6) _____ . Another necessary position on these large incidents is called the logistics section. This section will be responsible for securing (7) _____ and (8) _____ for the incident. Lastly the (9) _____ will be appointed in order to keep costs and timekeeping forms for the incident so it does not become unmanageable. To round out the system the incident commander will have to appoint a safety, information, and liaison officer to complete the (10) _____ . To get the necessary resources to an incident, departments may rely on (11) _____ with other jurisdictions.

12. Identify some of the specialty positions available in the fire service as shown in Figure 2-4.

    a. _____

    b. _____

    c. _____

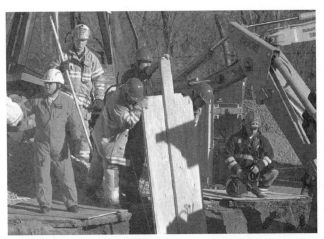

a.

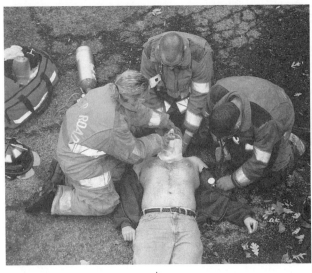

b.

c.

**Figure 2-4** Specialty positions available to firefighters.

# CHAPTER
# 3

# COMMUNICATIONS AND ALARMS

Photo courtesy of Scot Smith, Smith Photographic, Shreveport, Louisiana

# QUESTIONS

## Matching

Match the correct term with the definitions provided.

a. personal pager

b. location

c. communications center

d. mobile data terminal (MDT)

e. enhanced 9-1-1

f. power failure

g. authorized personnel

h. manual run card system

i. deaf

j. municipal fire alarm system

1. A(n) _____ may also be referred to in some instances as a "public safety answering point."

2. The only people who should be allowed into the communications center are _____ because of security reasons.

3. In case of a(n) _____, battery power may be used for critical communications and computer equipment in the dispatch center.

4. According to the text, when collecting incident information, the call taker must put more emphasis on understanding the incident's _____ than anything else.

5. The benefit of _____ is that the communications center is provided with the telephone number and address of the caller electronically.

6. A highly visible, easily recognizable, and highly accessible alarm system for the public's use is termed a(n) _____ .

7. Telecommunications devices for the _____ (or TDDs) are becoming much more widely used.

8. A card file containing street and location information for the dispatcher is available to agencies not using computers, and the system is called a(n) _____ .

9. Volunteer firefighters may receive alarms in a number of ways. The most typical is the _____ , but some also use the telephone system in the volunteer's house.

10. One of the newer innovations allowing dispatch information to be transmitted directly to the apparatus is the _____ , where the information is displayed on a screen in the cab.

## True or False

1. The primary role of the telecommunicator is to give the public information to dispatch.

   a. true     b. false

2. In some cases the telecommunicator may serve as the public relations officer for the department.

   a. true     b. false

3. It is recommended that even the slowest departments with very low call volumes have at least three people on duty in dispatch at all times.

   a. true     b. false

4. Modern emergency communications facilities are typically constructed in areas where there is little risk of natural or man-made hazards.
   a. true     b. false

5. If the communications facility is up to code and properly staffed, it is not important that the facility have a backup location.
   a. true     b. false

6. Because of today's electronic systems, it is not important that the call taker speak slowly or clearly, since the system is designed to enhance the communications.
   a. true     b. false

7. The citizen is the "customer," and in most successful businesses, fire departments included, the customer is always right. The call taker should never argue with a caller.
   a. true     b. false

8. Incomplete calls are usually pranks, so busy dispatchers should not work very long at following up on those calls. They must focus on full calls for assistance.
   a. true     b. false

9. When taking emergency medical calls, the telecommunicator may provide medical assistance using predetermined actions while units are en route.
   a. true     b. false

10. Cellular phone technology has evolved so far as to predetermine where the call has originated and then route the call to the nearest fire station.
    a. true     b. false

11. The Americans with Disabilities Act entitles citizens to equal service from public agencies.
    a. true     b. false

12. A still alarm is a call in which the victim or informant walks into the local emergency facility and calls for aid for himself or another.
    a. true     b. false

13. The most important information one can take on a call is the informant's name, since this ensures accuracy and discourages false alarms.
    a. true     b. false

14. One of the more important features of the CAD system is that predeployment factors are written into the system for quicker response.
    a. true     b. false

15. The newer "trunked" radio systems are a great benefit to field personnel because there is virtually no need to delay speaking when keying the microphone.
    a. true     b. false

16. Firefighters should not touch any radio antenna during transmission to avoid burns that can result from radio frequency energy.
    a. true     b. false

# Multiple Choice

1. Which of the following is *not* considered one of the four basic elements of the communications process?
   a. receiving information
   b. understanding information
   c. recording information
   d. reporting information

2. Telecommunicators should maintain contact with the caller in order to obtain valuable _____ for first responders:
   a. prearrival instructions
   b. name and return number
   c. medicine allergies
   d. insurance information

3. The NFPA Standard for Communications Centers states that 95 percent of alarms shall be answered within how many seconds, minutes, or rings?
   a. ten seconds
   b. thirty seconds
   c. one minute
   d. three rings

4. Where would the most desirable location for a communications center be?
   a. high traffic areas
   b. meadows
   c. nonrisk areas
   d. in collapse zones of other structures

5. The use of computers for communications centers is a big benefit. One important feature of computers in this area would be:
   a. resource tracking
   b. mathematics work
   c. quick issue investigation
   d. spell check

6. Incoming telephone calls to dispatch should be answered in the following priority:
   a. 9-1-1, business calls, personal calls
   b. emergency calls, personal calls, business calls
   c. 9-1-1 calls, direct lines, business calls
   d. direct lines, emergency calls, business calls

7. A call taker should always attempt to obtain the following information, with which answer being the least important?
   a. location of the emergency
   b. nature of the emergency
   c. callback number
   d. size of the fire

8. A(n) _____ is an alarm system operating in a protected premises.

   a. local protective signaling system

   b. remote station signaling system

   c. central station protective signaling system

   d. auxiliary protective signaling system

9. Of the following, which is not a method of receiving a report of an emergency?

   a. conventional telephone

   b. call box

   c. still alarm

   d. smoke detector

10. A(n) _____ is a device that converts an entered code into paging codes, which activate a number of paging devices.

   a. TDD

   b. ringdown circuit

   c. encoder

   d. CAD system

11. Of the following types of calls received by a communications center, for which one are call takers least likely to know the location?

   a. box alarm

   b. still alarm

   c. cell phone

   d. fixed site

12. In the mid-1800s, emergency call boxes were first installed in the United States. Which of the following cities claims to have been the first to bring them in and still uses them to this day?

   a. Boston

   b. San Francisco

   c. New York

   d. Los Angeles

13. An alarm system that connects a protected premise to a privately owned monitoring site and monitors the lines constantly is called a:

   a. heat-actuated system

   b. wet pipe system

   c. central station signaling system

   d. security alert system

14. Initiating the signal of a fire alarm system in an occupancy is most usually accomplished by all but which of the following?

   a. manual pull boxes

   b. heat detectors

   c. smoke detectors

   d. cell phones

15. According to the text, what are the primary items employed in a deployment plan utilized by telecommunicators?

    a. apparatus types, number of personnel, location of alarm

    b. nature of the call, equipment carried on units, number of personnel

    c. apparatus types, number of personnel, skill levels

    d. nature of the call, location of the call, equipment needed

16. One of the main regulations of the Federal Communications Commission (FCC) that greatly affects the fire service is:

    a. time allowed on the air

    b. frequency allocation

    c. cost of airtime

    d. radio distribution

17. One of the main reasons for using tones on the radios is for:

    a. safety

    b. covering unnecessary traffic

    c. indicating frequency change

    d. checking frequency range

18. Besides the use of radios to evacuate a structure in an emergency, many departments are utilizing:

    a. runners

    b. police

    c. air horns

    d. PAL devices

19. When arriving on scene, the responding unit should give a report over the radio to dispatch telling what they have found. That report should contain a number of items such as: (choose two)

    a. the correct address

    b. the number of people on the unit

    c. where in the building the emergency is located

    d. the color of the smoke

20. To assist on the fireground, some departments send out mobile support vehicles. The best definition of an MSV is:

    a. a police unit

    b. a communications unit

    c. a rehabilitation truck

    d. heavy rescue

21. A Type A alarm box system, as shown in Figure 3-1, is designed to dispatch to which of the following?

    a. guard station

    b. alarm company

    c. fire stations

    d. local police station

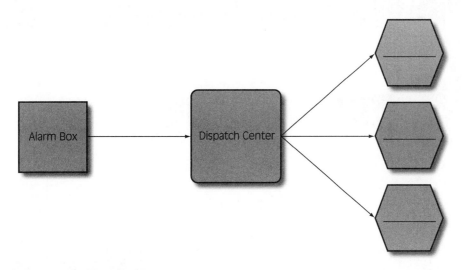

**Figure 3-1** Type A municipal alarm system.

# Fill-in-the-Blanks

    a. type B

    b. uninterruptible power supply (UPS)

    c. one; two; 45

    d. callback number

    e. 9-1-1 or other emergency lines

1. A(n) _____ is used in many communications centers to provide uninterrupted conditioned power to the critical communications and computer equipment in the event of a primary power failure.

2. _____ should be answered before all other incoming phone lines.

3. Getting and verifying the _____ of the caller reporting an emergency is important in the event it is necessary to recontact the caller for additional information.

4. _____ municipal alarm systems typically transmit alarms directly from a call box to first responders.

5. When transmitting on mobile or portable radios, hold the microphone _____ to _____ inches from the mouth at a _____ degree angle.

# FIRE BEHAVIOR

# QUESTIONS

## Matching

Match the correct term with the definitions provided.

  a. British thermal unit (Btu)
  b. exothermic
  c. combustion
  d. pyrolysis
  e. endothermic
  f. thermal layering

1. The term used for rapid oxidation with the development of heat and light is most usually
   _____ .

2. The temperature of a fire in fire department terms is most usually in _____ , and this
   term is used when figuring how much water to use in extinguishment.

3. When heat is given off during the reaction process, as with dynamite, the reaction is called
   _____ .

4. When heat is absorbed in the creation of a chemical bond, the reaction is called
   _____ .

5. Most organic solids decompose, giving off flammable gases when burning. This process is called
   _____ .

6. The stratification of gases produced by fire into layers based on their temperature
   is_____ .

7. Match the extinguisher classification letters to the shapes in Figure 4-1 as they are presented on
   extinguishers: A, B, C and D.

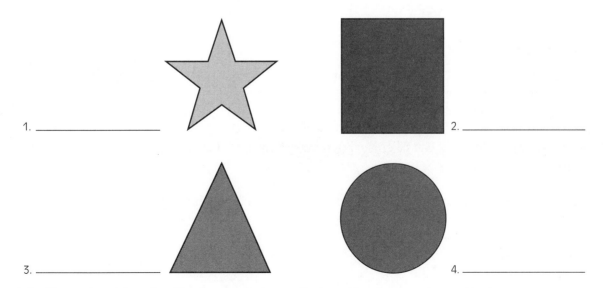

1. _____

2. _____

3. _____

4. _____

**Figure 4-1**   Fire extinguisher classification symbols are displayed by shape, color, and/or letter.

8. The objects in Figure 4-2 represent some basic firefighter knowledge. Fill in the blanks with the terms below:

heat

fuel

fire tetrahedron

heat

chemical reaction

fire triangle

fuel

oxygen

oxygen

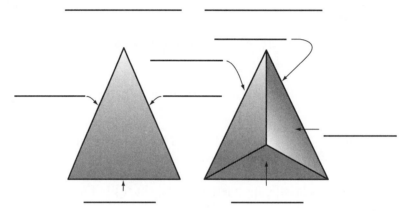

**Figure 4-2** Fill in the blanks for these two items.

# True or False

1. The ability to recognize a chemical compound will many times assist the firefighter in determining what extinguishing agent to apply.
   a. true        b. false

2. As an energy source, heat can be created or destroyed at any time given the proper tools.
   a. true        b. false

3. Of the four sources of heat, chemical is the most common in firefighting.
   a. true        b. false

4. The most common cause of heat buildup in machinery is electrical.
   a. true        b. false

5. A static charge can build up temperatures of more than 2,000°F.
   a. true        b. false

6. In a closed atmosphere, as oxygen decreases, the buildup of carbon monoxide and heat increases.
   a. true      b. false

7. Knowing that a gas is heavier than air will assist firefighters, since the gases will tend to dissipate quickly into the atmosphere.
   a. true      b. false

8. The more molecules crushed into a given volume, the denser the substance will be.
   a. true      b. false

9. During a BLEVE the release of gas and liquid into the air will create a fireball if the material is flammable.
   a. true      b. false

10. If a concentration of gas is below its flammable or explosive range, then it is too rich to burn.
   a. true      b. false

11. A flashover is a sudden event that occurs when all the contents of a container reach their ignition temperature nearly simultaneously.
   a. true      b. false

# Multiple Choice

1. The fire tetrahedron is said to consist of all but which of the following choices?
   a. fuel
   b. smoke
   c. chemical reaction
   d. heat

2. According to the text, iron, sulfur, and granite are examples of:
   a. oxides
   b. molecules
   c. inorganic substances
   d. endothermic reactions

3. Rusting iron is an example of which process?
   a. heat sink
   b. moist air
   c. conflagration
   d. oxidation

4. Sources of heat include all but which of the following?
   a. decombustion
   b. electrical
   c. nuclear
   d. chemical

5. A static charge can be dangerous if one of the following is present during the discharge.

    a. oxidation

    b. flammable gas

    c. friction

    d. wet carpet

6. Which of the following best describes a self-feeding chemical reaction that releases heat energy and reinvests it into the reaction?

    a. Btu

    b. BLEVE

    c. combustion

    d. oxygenation

7. What occurs when the chemical reaction of rusting takes place?

    a. moisture is diffused

    b. chemistry is reduced

    c. inorganics are formed

    d. oxidation takes place

8. When moisture is expelled from a liquid, molecules escape and return to the surface at different rates. This is called:

    a. evaporation

    b. boiling

    c. oxidation

    d. radiation

9. When figuring vapor density, normal air is given the weight (value) of:

    a. 0

    b. 1

    c. 5

    d. 12

10. A metallic vessel containing great pressure and heat ruptures, possibly causing which of the following?

    a. decompression

    b. BLEVE

    c. oxidation

    d. pyrolysis

11. All materials will obey the laws of nature and will exist in all but which of the following states?

    a. solid

    b. gas

    c. California

    d. liquid

12. The stages of fire include which two of the following? (choose two)
    a. behavior stage
    b. detection stage
    c. ignition stage
    d. decay stage

13. Which stage of a fire is recognized as the point where all of the contents within the enclosure are burning?
    a. behavior stage
    b. detection stage
    c. fully developed stage
    d. decay stage

14. Which of the following modes of heat transfer is not recognized as a true mode?
    a. radiation
    b. convection
    c. conduction
    d. ignition

15. The ability of a material to conduct thermal energy depends on that material's:
    a. space
    b. temperature
    c. organics
    d. density

16. In a burning liquid the ability to burn is dependent upon the substance's ability to place its molecules into suspension. Therefore, a liquid cannot burn unless it is in:
    a. evolution
    b. suspension
    c. flux
    d. oxidation

17. Which two of the following best describe methods of fire extinguishment? (choose two)
    a. thermal layering
    b. oxygen elimination
    c. triangulation
    d. chemical flame repression

18. Which of the following best describes a Class A fire?
    a. flammable liquids
    b. burning magnesium
    c. ordinary combustibles
    d. racing fuel

19. The Class A fire is best extinguished by which of the following methods?

    a. removal of the heat

    b. shutting off the power

    c. closing the valve

    d. application of graphite

20. Which of the following methods usually extinguishes the Class B fire?

    a. propagating the chain reaction

    b. removing the water

    c. keeping oxygen from the fuel

    d. sweeping with water

21. When the gases produced by fire stratify into layers based on their temperatures, what occurs?

    a. thermal conductivity

    b. thermal blanketing

    c. flashover

    d. thermal layering

22. The hottest gases produced by fire are located where in a closed container?

    a. near the top of the container

    b. near the middle of the container

    c. near the bottom of the container

    d. mixed throughout the container

23. When the entire contents of a container, or room, reach their respective ignition temperatures, what can occur?

    a. backdraft

    b. thermal layering

    c. banking down

    d. flashover

# Fill-in-the-Blanks

    a. radiated

    b. BLEVE

    c. Class A

    d. atomization

    e. explosive range

    f. exothermic

    g. heat

    h. endothermic

A closed container holding a flammable liquid is heated until it ruptures. As the gases are released they undergo (1) _____ . This (2) _____ can be disastrous and as the liquid is released into the air, it burns as its (3) _____ is reached, causing a large fireball. The (4) _____ heat will also cause many other combustibles in the area to ignite. What was once a Class B fire will now become a number of (5) _____ fires.

The two reactions in Figure 4-3 differ because the (6) _____ transfer differs. The top reaction is known as a(n) (7) _____ reaction, and the bottom figure represents a(n) (8) _____ reaction.

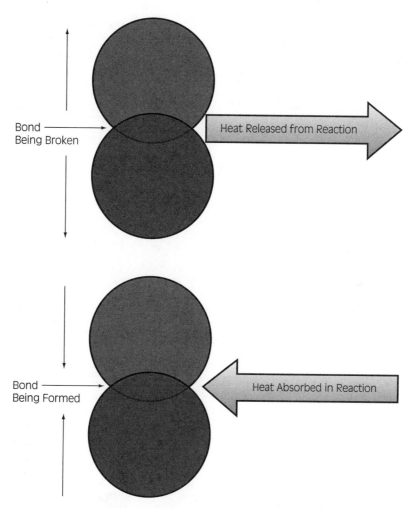

**Figure 4-3** A chemical reaction sequence.

9. Identify the form of heat demonstrated in Figure 4-4. _____

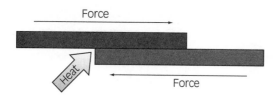

**Figure 4-4** Form of heat.

10. Fill in the correct vapor densities shown in Figure 4-5 using "greater than 1" and "less than 1."

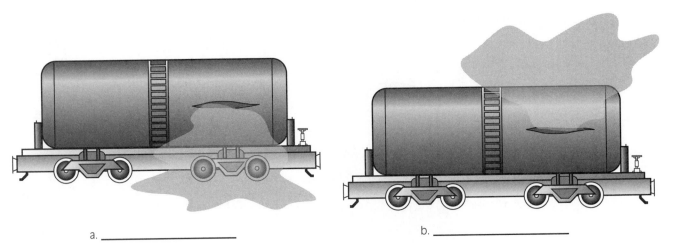

a. _____

b. _____

**Figure 4-5**   Identify each vapor density form.

11. Identify the four stages of fire as shown in Figure 4-6.

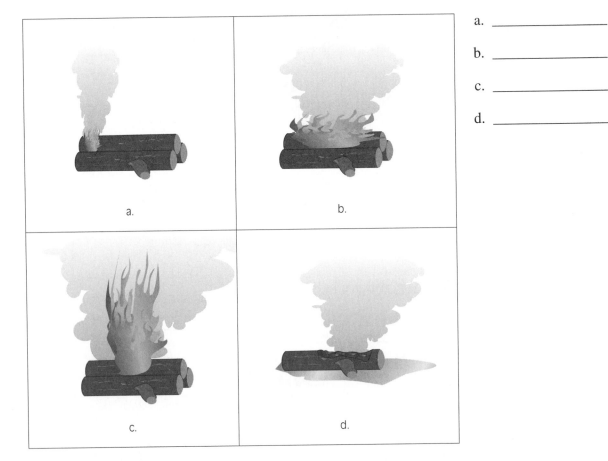

a. _____

b. _____

c. _____

d. _____

a.

b.

c.

d.

**Figure 4-6**   Identify the four stages of fire.

# FIREFIGHTER SAFETY

# QUESTIONS

## Matching

Match the correct term with the definitions provided.

a. accident        b. health and safety

c. employee assistance program        d. OSHA

e. intervention        f. personnel

g. procedures        h. equipment

1. The organization most responsible for safety code enforcement in the United States is _____ .

2. A(n) _____ can be defined as the result of events and conditions that led to an unsafe situation that resulted in injury and/or property damage.

3. _____ is said to have occurred when any action designed to break the accident chain takes place.

4. Employee drug and alcohol dependence, depression, and worker relationships as well as job stress are some of the issues that can be addressed through an organization's _____ .

5. Many departments are developing _____ committees made up of personnel from all levels to combat unsafe conditions and actions.

6. The safety triad is represented by:

     a. _____

     b. _____

     c. _____

## True or False

1. To keep firefighter injury and death events at a minimum, one must understand what events typically lead to accidents.
   a. true      b. false

2. Study of injury and death is beneficial, but will not prevent additional accidents.
   a. true      b. false

3. Data has shown that firefighter injuries are decreasing due to safety programs, while deaths, unfortunately, are on the rise.
   a. true      b. false

4. OSHA has had a tendency to follow the lead of the NFPA in addressing issues of firefighter safety.
   a. true      b. false

5. Intervention is typically a reactive action, while mitigation is typically a proactive action.
   a. true      b. false

6. Only formal procedures are responsible for department safety, while informal procedures usually are not.
   a. true      b. false

7. It is widely accepted that once people enter the fire service their training never ends.
   a. true      b. false

8. When addressing factors that affect safety, attitude is the easiest to work with.
   a. true      b. false

9. There is no better time in a person's career to positively shape a safety attitude. Whether early in the career or later, it makes no difference.
   a. true      b. false

10. The separation of members within a team is in many incidents a contributing factor to firefighter fatalities.
    a. true      b. false

## Multiple Choice

1. The unfortunate deaths of a number of firefighters in Hackensack, New Jersey, were eventually blamed mainly on which of the following?
   a. poor water supply
   b. bowstring trussed roof
   c. not enough manpower
   d. a BLEVE

2. The studies of firefighter injury and death statistics have shown that about how many of all on-duty deaths and injuries occur at the incident scene?
   a. 10 percent
   b. 25 percent
   c. 50 percent
   d. 75 percent

3. What is the leading type of death-causing injury?
   a. roof collapse
   b. wall collapse
   c. burnover
   d. heart attack

4. The Code of Federal Regulations dictates specific processes and procedures to be followed by both the fire service and private business in safety compliance. What agency enforces that code?
   a. OSHA
   b. NFPA
   c. YMCA
   d. EDITH

5. The standard that is dedicated to firefighter occupational health and safety is termed:
   a. NFPA 1021
   b. NFPA 1500
   c. ICS 300
   d. EMT 1A

6. Which of the following items is not one of the five components of the accident chain?

   a. environment

   b. human factors

   c. injury relationships

   d. equipment

7. According to the text, which of the following will most ultimately protect firefighters from injury?

   a. good equipment

   b. strong leadership

   c. great stamina

   d. positive attitude

8. The safety triad consists of all but which of the following?

   a. accidents

   b. procedures

   c. equipment

   d. personnel

9. It is said that most critical equipment used in firefighting is designed and built to meet which of the following?

   a. IAFF laws

   b. OSHA regulations

   c. IBC codes

   d. NFPA standards

10. According to the text, without a doubt, the single most important step that can be taken to reduce firefighter injuries is that of:

    a. better materials in clothing

    b. longer duration SCBA

    c. physical strength

    d. regular training

11. Of the following choices, which is *not* one of the keys to improving physical health and therefore firefighter safety?

    a. annual health screening

    b. firefighter nutrition education

    c. mandatory fitness programs

    d. mandated safety clothing

12. As much as physical health is important in firefighting, so is firefighter mental health. Therefore, which of the following systems has been developed for better mental health?

    a. AAA

    b. CISM

    c. OSHA

    d. NFPA

13. What is the purpose behind employee assistance programs?

    a. employee and family assistance

    b. nutritional guidance

    c. physical agility

    d. promotional ability

14. If a department is going to create a health and safety committee for injury and accident prevention, the makeup of the committee should be from which of the following?

    a. the administration

    b. the floor

    c. outside the department

    d. all levels

15. A firefighter's readiness includes not only wearing PPE and SCBA as appropriate, but also being ready: (choose two)

    a. every day

    b. mentally

    c. for many years

    d. physically

16. Figure 5-1 establishes the sequence for accidents. What is this sequence called?

    a. the injury sequence

    b. risk stages

    c. the accident chain

    d. event factors

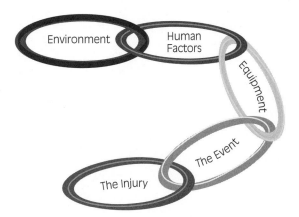

**Figure 5-1** Identify this figure.

17. The individual firefighter contributes to the safety partnership by performing all of the following *except:*
    a. understanding the chain of command
    b. freelancing
    c. performing as trained
    d. being ready

18. To help achieve a standard approach to handling incidents, firefighters should use:
    a. a set of pocket tools
    b. nationally recognized apparatus
    c. an incident engagement checklist
    d. standard freelance guidelines

19. Work hardening is the effort to:
    a. improve immunization offerings
    b. make nutritional lifestyle changes
    c. set positive examples by line officers and working firefighters
    d. better perform physical tasks without overstressing

# CHAPTER 6

# PERSONAL PROTECTIVE CLOTHING AND ENSEMBLES

# QUESTIONS

## True or False

1. An important point to remember is that PPE is the firefighter's first line of protection and that it must be worn properly.

   a. true      b. false

2. It is the manufacturer's responsibility to verify that garments meet or exceed NFPA standards by attaching a label in the garments.

   a. true      b. false

3. Failure to use PPE in certain conditions (i.e., IDLH atmospheres) is a violation of federal regulations.

   a. true      b. false

4. Some departments still use turnout coats and three-quarter pull-up boots. NFPA allows this only if the boots are overlapped 6 inches by the coat.

   a. true      b. false

5. It is important to the firefighter that the structural firefighting hood carries the same TPP rating as the other gear so that the ensemble carries the proper ratings.

   a. true      b. false

6. Fighting a wildland fire with structural PPE is inviting an injury because of the different demands of the work.

   a. true      b. false

7. The concept behind the PASS device is that motion will set off an alarm, calling attention to the firefighter in distress.

   a. true      b. false

8. A work uniform that meets NFPA standards is designed to protect the wearer from IDLH atmospheres.

   a. true      b. false

## Multiple Choice

1. According to the text, proper fire streams, zoning, and sound tactics should accomplish which of the following in regard to safety?

   a. achieve quick response

   b. guarantee success on the scene

   c. define proper protocol for arriving units

   d. provide a greater measure of safety

2. Which of the following is most usually considered firefighter PPE? (choose two)

   a. gloves

   b. hose

   c. hoods

   d. pike poles

3. Which of the following is considered a result of the evolution of materials and sound risk management practices?
   a. fireground improvements
   b. improvements in protective clothing
   c. faster response
   d. firefighting hoods

4. The NFPA has developed minimum ensemble standards for all but which of the following?
   a. all risk environments
   b. structural firefighting clothing
   c. wildland firefighting clothing
   d. proximity PPE

5. Which of the following items is not part of the layered structural firefighting coat?
   a. heat-resistant lining
   b. vapor barrier
   c. outer shell
   d. thermal barrier

6. What is the definition for TPP when used in conjunction with safety clothing?
   a. true temper preparation
   b. protection from abrasive contact
   c. protection time before wearer injury
   d. training to perfection

7. Gloves must accomplish a number of things. Name two things that they are tested for (choose two).
   a. puncture resistance
   b. thermal resistance
   c. water retention
   d. colorfastness

8. According to the text, an important interface that creates an encapsulating link to the firefighter's helmet, coat, and SCBA is which of the following?
   a. PPE mandates
   b. turned-up collar
   c. face shield
   d. protective hood

9. The proximity firefighting PPE ensemble is most often associated with which of the following?
   a. hazardous materials response
   b. aircraft rescue and firefighting
   c. collapse structure response
   d. medical emergencies

10. Layering is important in all forms of firefighting clothing. What item is used to layer the wildland ensemble?

    a. pull-up boots

    b. vapor lining

    c. cotton undergarments

    d. polyester underwear

11. What is the firefighter's last resort for safety in the wildland environment when all else fails?

    a. physical stamina

    b. superb training

    c. aircraft rescue

    d. fire shelter

12. What is the item called that carries the canteen, radio, flares, and fire shelter?

    a. harness

    b. backpack

    c. utility belt

    d. web gear

13. What is the term given for the motion detection device worn by firefighters?

    a. SCBA

    b. PASS

    c. hood

    d. low-pressure warning

14. The NFPA recommends cleaning safety clothing how often?

    a. every six months

    b. every twelve months

    c. every eighteen months

    d. every twenty-four months

15. What is meant by acclimation regarding safety clothing?

    a. getting used to the gear

    b. altitude adjustment

    c. proper fit

    d. cleaning practices

16. The three items shown in Figure 6-1 are basically used for the same purpose. What is that purpose?

    a. regulate air pressure

    b. low air warning

    c. lost personnel identity

    d. air capacity of unit

17. A PPE ensemble that includes a buoyant, insulated suit can best be described as a(n):

    a. ice rescue ensemble

    b. swift water ensemble

    c. technical rescue ensemble

    d. arctic ensemble

18. To help beat the discomfort of wearing PPE, firefighters should:
    a. develop shortcuts
    b. define temperature ranges suitable for each PPE ensemble
    c. participate in on-scene rehab
    d. dress down at every opportunity

19. Mastery is the concept that an individual can achieve:
    a. proficiency in skills
    b. 70% of an objective 70% of the time
    c. perfection in skills
    d. 90% of an objective 90% of the time

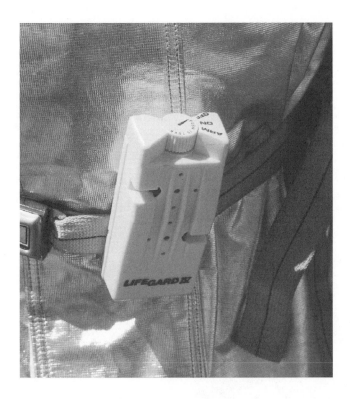

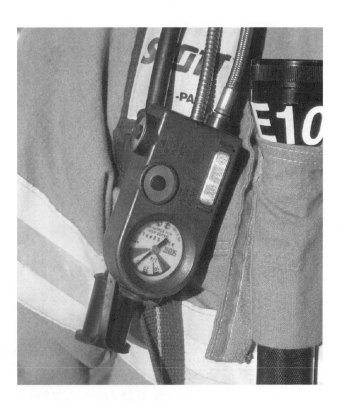

**Figure 6-1** What is the purpose for these items?

## Fill-in-the-Blanks

    a. NFPA

    b. ANSI

    c. OSHA

    d. IAFF

    e. CDC

    f. EPA

    g. NIOSH

    h. ASTM

Name five of the organizations associated with the safety and development of PPE using the preceding acronyms.

    1. _____

    2. _____

    3. _____

    4. _____

    5. _____

    6. When preparing to enter the fireground, personnel should go through the system shown in Figure 6-2. What is this system called? _____

**Figure 6-2** Name this two-person system.

# SELF-CONTAINED BREATHING APPARATUS

# QUESTIONS

## Matching

Match the correct term with the definitions provided.

    a. carbon monoxide

    b. PASS

    c. mask rule

    d. respiratory protection

    e. smoke

    f. SCBA regulator

1. The _____ should require all personnel to not only wear but use their SCBA during operations where IDLH is present.

2. An audible warning device that incorporates a motion detector and automatically sounds if movement is not sensed within thirty seconds is termed a(n) _____ and is mandated in IDLH conditions.

3. The combination of unburned products of combustion is most often termed _____ and is present in most structure fire IDLH conditions.

4. _____ is a colorless and odorless gas that is almost always present in fires and is a deadly ingredient of an IDLH condition.

5. Both 29 CFR 1910.134 and NFPA 1500 contain similar requirements for _____ .

6. Figure 7-1 shows two versions of the same thing. What are the items in this figure? _____

**Figure 7-1** Identify these items.

# True or False

1. The weight of the standard thirty-minute SCBA used in the fire service has been slightly increased in the past twenty years because more air is carried in the units.

   a. true     b. false

2. With a positive pressure SCBA an airflow slightly above atmospheric pressure is maintained in the mask.

   a. true     b. false

3. All of today's fire service SCBAs are, by law, the positive pressure design.

   a. true     b. false

4. Fire consumes smoke and produces toxic gases that may displace carbon monoxide in the burning environment.

   a. true     b. false

5. Fire extinguishing systems such as total flooding $CO_2$ will displace oxygen, thus requiring the use of SCBA in those environments.

   a. true     b. false

6. Temperatures in a structure fire can reach more than 1,000°F, and a firefighter can only last five minutes in that environment.

   a. true     b. false

7. After a fire is extinguished, the carbon monoxide levels will immediately drop and the firefighter can remove the SCBA.

   a. true     b. false

8. Carbon monoxide will combine with hemoglobin in blood somewhat more slowly than oxygen, so firefighters must wear SCBA whenever exposed to oxygen.

   a. true     b. false

9. An increased respiratory rate combined with high CO levels will incapacitate a firefighter very quickly if SCBA is not worn.

   a. true     b. false

10. Depending on the activity level and physical condition of the wearer, an SCBA cylinder with a rating of thirty minutes may last considerably less than thirty minutes.

    a. true     b. false

11. Hydrogen cyanide is a deadly gas and has the distinct odor of peanuts.

    a. true     b. false

12. CO exposure symptoms range from a mild headache to death.

    a. true     b. false

13. After attaching the regulator, a firefighter should inhale and hold a breath for about 5 seconds and listen and feel for any leaks.

    a. true     b. false

## Multiple Choice

1. According to the text, failure to properly use SCBA could result in all but which of the following?
   a. failure to rescue victims
   b. injury or death of the firefighter
   c. long-term health effects to the firefighter
   d. improper use of fire stream

2. Many changes have occurred in SCBA over the years. Which two of the following are proof of these changes? (choose two)
   a. SCBA weight reduction
   b. change from oxygen to air
   c. less use of air demanded
   d. positive pressure apparatus

3. Programs have been developed to increase firefighter knowledge and confidence in using SCBA. Which of the following is a good example of one of these programs?
   a. PASS program
   b. cascade stations
   c. smoke divers
   d. pressure demand

4. Which of the following is not considered a condition that presents respiratory hazards commonly found at fire or other emergency incidents?
   a. venting of gases
   b. oxygen deficiency
   c. high temperatures
   d. toxic environments

5. Ordinarily about 98 percent of the oxygen breathed in is carried by which part of the blood?
   a. riboflavin
   b. monoxide
   c. hemoglobin
   d. air cells

6. The primary mandatory requirement concerning the use of SCBA is established by which legal authority?
   a. NFPA
   b. OSHA
   c. IAFF
   d. EDITH

7. SCBA has a number of limitations in its use. Which of the following limitations is *not* one listed in the text?
   a. room temperatures
   b. weight of unit
   c. limited air supply
   d. size of unit

8. Which two of the following types of SCBA are most common in today's fire service? (choose two)
   a. open circuit
   b. closed circuit
   c. outside ventilated
   d. canister

9. Which two of the following standards organizations have the most input into the development of today's SCBA? (choose two)
   a. OSHA
   b. NIOSH
   c. CHEMTREC
   d. NFPA

10. When considering the basic components of an SCBA system, all but which of the following is included in this description?
    a. backpack and harness
    b. cylinder
    c. regulator
    d. canister

11. Most of the SCBA units today are designed so that the wearer carries most of the weight on what part of the body?
    a. shoulders
    b. hips
    c. arms
    d. legs

12. Just as with all PPE, SCBA should be kept clean. The best solution for washing composite bottles and harnesses is usually:
    a. mild soap and water
    b. laundry detergent
    c. scouring powder
    d. chemical cleansers

13. The compressed air used to fill SCBA is rated by a letter. What is the minimum grade of air accepted in the refilling of these cylinders?
    a. A
    b. B
    c. C
    d. D

14. According to OSHA and NFPA regulations, the supply source for breathing air must be tested for air purity how often?
    a. every three months
    b. every six months
    c. annually
    d. biannually

15. There are two pressure gauges on every SCBA unit. What is an allowable difference between these gauges?

    a. 150 lb
    b. 200 lb
    c. 500 lb
    d. no difference

16. What is being inspected in Figure 7-2?

    a. high pressure line
    b. regulator connect
    c. exhalation valve
    d. O-ring

**Figure 7-2** What is being inspected here?

## Fill-in-the-Blanks

    a. face piece
    b. hydrostatic
    c. aluminum
    d. SCBA
    e. carbon monoxide
    f. IDLH
    g. steel
    h. fiberglass
    i. regulator
    j. alarm

During repeated exposures to low levels of (1) _____ during a number of fires, a buildup will occur in the blood, causing a poisoning of the system. Because of this, OSHA has defined a hostile interior environment as a(n) (2) _____ atmosphere. The fire service uses (3) _____ to prevent injury in these environments.

Self-contained breathing apparatus consists of a number of components. The cylinder contains breathing air for the firefighter. That air is contained in cylinders made of (4) _____ , (5) _____ , and (6) _____ . These cylinders must undergo pressure tests on a regular basis called

(7) _____ tests. Another component of the SCBA assembly is called the
(8) _____ , and its purpose is to reduce the high pressure of the cylinder to a lower pressure sent
to the (9) _____ . If the pressure in the cylinder falls to approximately one-fourth of the rated
capacity of the unit a(n) (10) _____ will sound, giving the wearer warning of danger.

11. Name the four major parts of the SCBA in Figure 7-3.

4. _____

3. _____          2. _____     1. _____

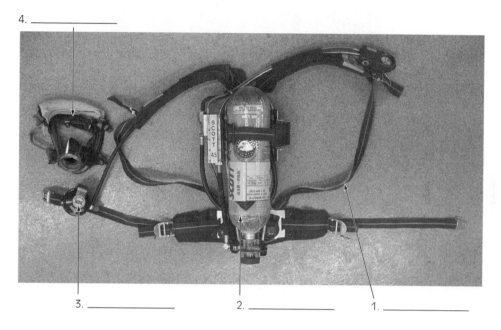

**Figure 7-3**   A typical SCBA with its components spread out for viewing.

12. What would make the occupancy in Figure 7-4 hazardous for firefighters if it were burning?

_____

**Figure 7-4**   What hazards can be expected here?

13. Figure 7-5 shows a high pressure hose being disconnected from an air bottle. If this connection is unusually tight it is an indication that what problem may exist? _____

_____

**Figure 7-5** Disconnect the high pressure coupling from the cylinder.

# PORTABLE FIRE EXTINGUISHERS

# QUESTIONS

## Matching

Match the correct term with the definitions provided.

        a. wet chemicals
        b. $CO_2$
        c. Class D
        d. Class A
        e. Class C

1. _____ fires are typically extinguished with water.

2. Fires involving Class B flammables are typically extinguished with _____ .

3. Dry sand and other special agents are typically used to extinguish _____ fires.

4. The new class of fire is termed Class K, and the agents of its extinguishment are typically _____ .

5. _____ fires are most usually extinguished with carbon dioxide.

6. Match the names of the parts of the stored pressure water extinguisher in Figure 8-1.
    a. discharge lever               b. water or solution
    c. discharge hose and nozzle     d. carrying handle
    e. siphon tube                   f. pressure gauge
    g. anti-overflow tube

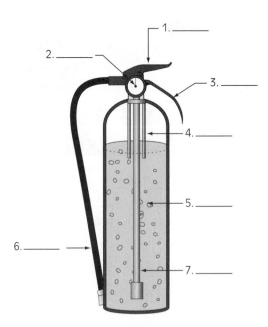

**Figure 8-1** Stored pressure water extinguisher.

7. Match the names of the parts of the carbon dioxide extinguisher in Figure 8-2.

    a. liquid

    b. discharge lever

    c. carrying handle

    d. gas

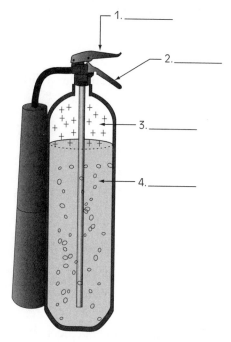

**Figure 8-2** Carbon dioxide extinguisher.

# True or False

1. It is not important to know the type of extinguisher carried on the engine since every unit carries both types.

    a. true      b. false

2. Pressurized flammable liquids and gases should not be extinguished if burning, unless the fuel supply can be shut off.

    a. true      b. false

3. Water is a good extinguishing agent for burning combustible metals since they could pose a hazard to the unwary firefighter.

    a. true      b. false

4. Class D agents are called dry powder agents, and they are very much like dry chemicals in their chemistry.

    a. true      b. false

5. Class E is a new classification that involves flammable cooking oils and fats.

    a. true      b. false

6. Temperature can be a factor in the placement of water-based extinguishing agents.

    a. true      b. false

7. Even though delicate equipment and high-value items may require special considerations in fire extinguishment, the main factor to consider is extinguishing the fire quickly and effectively.

    a. true     b. false

8. $CO_2$ is perfect for electrical fire work and virtually nonconductive.

    a. true     b. false

9. The best way to judge the level of material in an extinguisher is to check the gauge, which all extinguishers are required to have.

    a. true     b. false

10. The type of extinguisher as well as the size of the room has a direct bearing on the distance the firefighter must be from the fire to be effective.

    a. true     b. false

11. Using the wrong extinguisher is worse than using no extinguisher.

    a. true     b. false

12. Halon or halogenated hydrocarbon extinguishers are banned by an international treaty as these agents are thought to cause damage to the Earth's ozone layer.

    a. true     b. false

## Multiple Choice

1. When used by untrained people, fire extinguishers have been known to be: (choose two)
   a. practical
   b. quick
   c. ineffective
   d. dangerous

2. Of the following materials, which is the least practical in working a Class A fire?
   a. water
   b. foam
   c. dry chemicals
   d. soda ash

3. Gasoline, oils, alcohol, and propane are considered what type of fire?
   a. Class B
   b. Class C
   c. difficult
   d. none of the above

4. Common extinguishing agents for Class B fires are most usually: (choose two)
   a. water
   b. foam
   c. dry chemical
   d. wet sand

5. When extinguishing a Class C fire the best agents to use will usually be: (choose two)
   a. foam
   b. water
   c. $CO_2$
   d. dry chemical

6. Water-based solutions of potassium carbonate–based chemicals are good examples of extinguishing agents for which type of fire?
   a. Class A
   b. Class K
   c. Class D
   d. Class C

7. What type of fire would one be looking for when fixed extinguishing systems are used with flammable liquids?
   a. ordinary combustibles
   b. electrical
   c. deep-fat fryers
   d. paper goods

8. What is the first consideration in choosing the proper extinguisher? (choose two)
   a. type of fuels
   b. color of chemical in unit
   c. odor of agent
   d. amount of fuels

9. Name the material that is stored under pressure as a liquid that is capable of self-propelling from the extinguisher as a gas.
   a. dry chemical
   b. water
   c. carbon dioxide
   d. electrolyte

10. Dry chemicals are very effective due to which of the following ignition reduction actions?
    a. fuel reduction
    b. coating action
    c. cooling action
    d. wetting action

11. Which of the following types of extinguishers use an inert gas to propel the material out of the extinguisher? (choose two)
    a. dry chemical
    b. water
    c. foam
    d. $CO_2$

12 What is used in testing Class A extinguishers for extinguishing capacity?

    a. liquid pan

    b. smoke chamber

    c. wood cribbing

    d. paper bales

13. The test for the Class B extinguisher is typically which of the following?

    a. liquid pan

    b. wood cribbing

    c. smoke chamber

    d. not available

14. The testing of Class C extinguishers is typically done by which method?

    a. liquid pan

    b. wood cribbing

    c. paper bales

    d. not available

15. Of the following extinguishers, which is the only one still being used today?

    a. soda acid

    b. chemical foam

    c. carbon tetrachloride

    d. pressurized water

**Figure 8-3** Identify this extinguisher type.

16. Identify the extinguisher type shown in Figure 8-3.

    a. dry chemical

    b. water

    c. $CO_2$

    d. foam

17. Identify the extinguisher type shown in Figure 8-4.

    a. dry chemical

    b. water

    c. $CO_2$

    d. foam

**Figure 8-4** Identify this extinguisher type.

18. Which of the following indicates the proper sequence for the photos in Figure 8-5?

    a. a, b, c

    b. b, c, a

    c. b, a, c

    d. c, b, a

a.

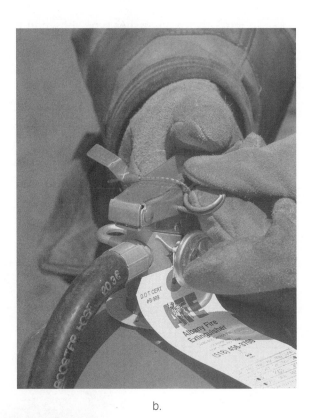

b.

c.

**Figure 8-5** Arrange this extinguisher sequence.

19. In 1998, a new class of fire was identified. This fire involves fires in combustible cooking oils, or animal oils and fats, and is known as:

a. Class E

b. Class F

c. Class H

d. Class K

# WATER SUPPLY

*Photo courtesy of Fred Schall*

# QUESTIONS

## Matching

Match the correct term with the definitions provided.

  a. water table

  b. fire flow

  c. gravity-fed

  d. groundwater

  e. cisterns

1. The amount of water required to put out a fire is termed _____ .

2. Water that seeps into the ground from rain and other surface sources is called _____ , and it is the primary source of supply.

3. Water stored underground tends to seek a level unique to the geographic area under study. This level is called a _____ , and it can change with seasonal rainfall and for other reasons.

▦ 4. Many times water is stored for future use in underground tanks called _____ , which are made out of rock or cement.

▦ 5. If the stored water is at a geographic level higher than the area in need, firefighters can just allow it to run down to them. This type of water supply is called _____ .

6. Using the following terms, match the hydrant to the types shown in Figure 9-1.

   a. wet barrel              b. dry barrel

1 _____          2 _____

**Figure 9-1**  Match the hydrant to its type.

7. A grid or looped system is represented in Figure 9-2. Identify the three main components of this system by placing the correct term in the space at the arrow.

    a. secondary feeders

    b. distributors

    c. primary feeders

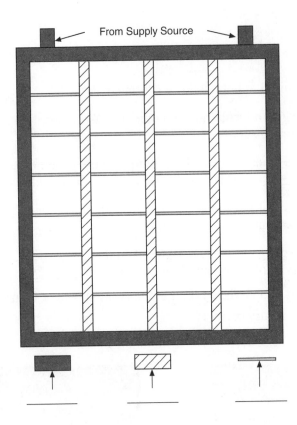

**Figure 9-2** Identify the three main components of this system.

8. Figure 9-3 shows three types of water distribution systems: combination gravity-pumped system, gravity-fed, and direct pump. Match the system types to the letter under the correct frame.

a. _____

b. _____

c. _____

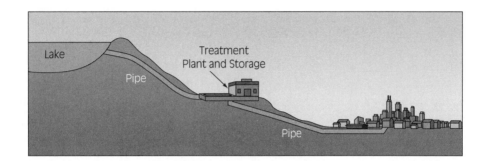

a.

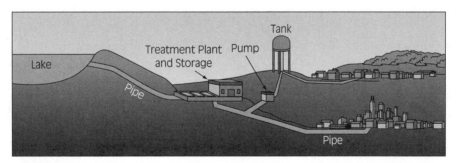

b.

c.

**Figure 9-3**  Water distribution systems.

# True or False

1. The most common extinguishing agent today is water.
   a. true    b. false

2. Water is important in areas where distribution systems are common, but not as important in rural areas where systems are not common.
   a. true    b. false

3. Most of the Earth's supply of fresh water comes from the ocean after it is filtered through the ocean bottom.
   a. true    b. false

4. In rural areas, the domestic and farm wells are a good source for firefighting because their pressure is fairly high.
   a. true    b. false

5. A water tender may come with a pump and hose, but it is also possible that it may not since there is no hard-and-fast rule.
   a. true    b. false

6. In many cases communities have used gasoline tankers for water tenders. Because they already carried liquid they do not need any alterations in order to make them efficient water tenders.
   a. true    b. false

7. The most common type of domestic water supply system is the combination pumped-gravity system in which part of the water is gravity-fed and the rest is pumped from wells.
   a. true    b. false

8. The order of water distribution in cities runs from the primary feeders to the distribution lines to the secondary feeders.
   a. true    b. false

9. Dead-end water mains are more reliable than other types because the pressure has to go only in one direction.
   a. true    b. false

10. Wet barrel hydrants are a form of hydrant in which water is brought into the barrel as soon as the valve at the top is opened.
    a. true    b. false

11. Connecting additional lines to a flowing wet barrel hydrant is no problem because each gate has its own valve.
    a. true    b. false

12. Firefighters must use caution when connecting to a wall hydrant because it differs from the fire pump test connection outlets.
    a. true    b. false

13. When using tenders as a water source in a rural setting, the best way to gain time is by asking the drivers to respond as fast as possible with the tenders, giving them better road speed.
    a. true    b. false

14. A partial vacuum created within the pump will cause water being drafted to enter the pump via a suction hose.
    a. true    b. false

15. Since hydrants are used so seldom, they should be tested periodically.
    a. true    b. false

## Multiple Choice

1. One of the main reasons for water being so effective in firefighting is its ability to:
   a. run downhill
   b. cover great areas
   c. be pumped easily
   d. absorb heat

2. Of the factors that affect water sources, which of the following has the greatest effect?
   a. day versus night
   b. tides
   c. geography
   d. weather

3. Water that enters the ground tends to collect in pockets called:
   a. pits
   b. tombs
   c. aquifers
   d. tanks

4. Which of the following is the most common source of water on the planet?
   a. clean
   b. surface
   c. rain
   d. fire

5. Most wells require what type of device to draw water to the surface for use?
   a. pump
   b. tank
   c. tender
   d. gas

6. Which of the following is *not* a good natural source of water?
   a. swimming pool
   b. river
   c. lake
   d. pond

7. Which of the following is a good man-made source of water?

   a. ocean

   b. stream

   c. reservoir

   d. lake

8. Water tanks may be underground, at ground level, or elevated, and some may also have: (select two best answers)

   a. drafting points

   b. dry hydrants

   c. fire department connections

   d. OS&Ys

9. When designing a water system, it is best to design it to meet the public needs at what time of day?

   a. night

   b. day

   c. peak

   d. does not matter

10. Water is supplied in all but which of the following ways?

    a. gravity-fed

    b. pumped

    c. combination gravity and pumped

    d. quick flow

11. Good water distribution systems are interconnected into: (choose two)

    a. wells

    b. grids

    c. loops

    d. ponds

12. When more than one pumper is connected to a water main, which of the following types of systems is the least dependable?

    a. loop

    b. grid

    c. pond

    d. dead-end

13. The two major hydrant types are: (choose two)

    a. wet barrel

    b. below ground

    c. dry barrel

    d. gravity

14. What type of hydrant is best suited for areas that have below-freezing temperatures?
    a. wet barrel
    b. belowground
    c. dry barrel
    d. gravity

15. Which type of hydrant uses a valve at the base to control water flow to all outlets?
    a. wet barrel
    b. belowground
    c. dry barrel
    d. gravity

16. For drafting, the best hydrant to use is the one with what device at the end in the water?
    a. strainer
    b. cap
    c. tube
    d. suction

17. What type of hydrant would one find on an airport runway?
    a. cistern
    b. gravity
    c. flush-type
    d. high profile

18. A backflow preventer is what type of valve that prevents backflow from system to system?
    a. check valve
    b. piston valve
    c. open valve
    d. gate valve

19. Moving water from source to need during an operation is best called:
    a. quick fill
    b. tanker response
    c. shuttle operation
    d. drafting

20. To speed the unloading of water tender tanks, some apparatus use a:
    a. jet-dump
    b. duck-foot
    c. pump-and-roll valve
    d. pressure surge preventer

21. A water column of 2.31 feet exerts a head pressure of:
    a. .433 psi
    b. 1 psi
    c. 14.7 psi
    d. 34 psi

22. A damaging, sudden surge of water caused by the quick opening or closing of valves is called a:
    a. tidal surge
    b. jet siphon
    c. backflow check
    d. water hammer

## Fill-in-the-Blanks

    a. water source
    b. dump valve
    c. dump site
    d. multiple tenders
    e. portable tanks
    f. pitot gauge

Regarding tender operations, the (1) _____ is where the water is delivered for quick unloading. Here a tender will arrive and set up (2) _____ for engine supply. The driver will pull up to the spot and operate a (3) _____ for filling the tank. If the operation requires a great amount of water, then (4) _____ will travel from the (5) _____ to the scene, unloading as quickly as possible.

What is the tool used in Figure 9-4 for testing the water pressure called? (6) _____

**Figure 9-4**  Hydrant testing.

# FIRE HOSE AND APPLIANCES

# QUESTIONS

## Matching

Match the correct term with the definitions provided.

- a. jacket
- b. forward lay
- c. fire hose
- d. booster line
- e. connect

1. The term _____ is used to define a flexible conduit used to convey water or other agents from a source to a fire.

2. Couplings, adapters, and appliances are used to _____ hose lengths together.

3. A _____ is applied to the hose at the factory in order to protect it from mechanical damage during its use.

4. The smaller hose carried on some apparatus is often called a _____ , and it is usually carried on a hose reel for quick but limited use.

5. When an engine stops at a water source, hooks up a supply line, then proceeds to the fire, the fire service calls that a _____ .

6. Using the following terms, match the type of hose roll to the photos in Figure 10-1.

   twin donut            straight            single donut

a._____

b._____

c._____

**Figure 10-1** Hose rolls.

# True or False

1. Today's fire service, for the most part, uses National Standard Thread. Those departments that do not use it will usually carry adapters.

   a. true     b. false

2. Most fire hose comes in lengths of 50 or 100 feet.

   a. true     b. false

3. Booster lines are a good, quick structural firefighting hose because they are so easy to maneuver.

   a. true     b. false

4. Attack hose is most often a minimum of 1½ inches to a maximum 4 inches in diameter.

   a. true     b. false

5. Wet hose should be dried prior to returning it to the apparatus with the exception of the newer synthetic hose, which can be reloaded wet.

   a. true     b. false

6. When passing over hose, a vehicle will do more damage if the hose is charged.

   a. true     b. false

7. A water hammer is possible if valves are opened or closed too quickly.

   a. true     b. false

8. Hose is really the most durable item on the engine. It only requires checking after freezing or exposure to flame.

   a. true     b. false

9. Hose couplings have lugs on them so that hose can be dragged. Pulling hose with the couplings on the ground is acceptable as long as the lugs are in good shape.

   a. true     b. false

10. If a knot is tied in the end of a hose when it is rolled, it is an indicator that the hose may be damaged.

    a. true     b. false

11. How much hose to lay on the fireground is determined by an officer who makes a judgment call on what is needed.

    a. true     b. false

12. When advancing line up a stairway the best method is to pull a charged line, since it must be ready at a second's notice.

    a. true     b. false

13. As opposed to a stairway pull, the ladder pull always pulls the hose uncharged because each section must be coupled when on the roof.

    a. true     b. false

14. If one had to choose the most practical between a forward and a backward hose lay, the backward is a better lay because it places the pressure end of the hose at the fire.

    a. true     b. false

15. With the forward lay the engine is left at the hydrant pumping the hose.

    a. true      b. false

## Multiple Choice

1. Fire hose is made of three types of construction. Which of the following is *not* one of the types?

    a. spiral

    b. wrapped

    c. braided

    d. woven

2. Name the type of hose that contains a wire helix to support the hose when drafting.

    a. booster line

    b. hard suction

    c. soft suction

    d. monitor line

3. Fire hose is generally divided into all but which of the following types?

    a. attack

    b. forestry

    c. chemical

    d. hard suction

4. Hose used regularly for fire attack is tested:

    a. weekly

    b. monthly

    c. every six months

    d. annually

5. Medium-diameter hose is usually which size?

    a. 1 in.

    b. 1¾ in.

    c. 2½ to 3 in.

    d. more than 3½ in.

6. Supply or LDH hose is usually which size?

    a. 1 in.

    b. 1¾ in.

    c. 2½ to 3 in.

    d. more than 3½ in.

7. Hard suction hose is standard in which length?

    a. 10 ft

    b. 20 ft

    c. 50 ft

    d. 100 ft

8. The main purpose of soft suction hose is for:
   a. drafting
   b. hydrant supply
   c. very long lays
   d. reverse lays

9. Soft suction hose generally runs in which of the following diameters?
   a. 2½ in.
   b. 3 in.
   c. 4½ to 6 in.
   d. 10 in.

10. The most common cause of hose damage on the fireground is usually:
    a. flame
    b. chemicals
    c. freezing
    d. coupling abuse

11. Hose couplings are divided into which two types? (choose two)
    a. American
    b. National
    c. unthreaded
    d. threaded

12. Which of the following materials is not commonly used in hose couplings?
    a. pyrolite
    b. steel
    c. brass
    d. aluminum

13. Two common ways of connecting hose couplings are: (choose two)
    a. leg raise
    b. over the hip
    c. foot tilt
    d. dutchman

14. If one finds that the hose has been rolled with the male end inside, it most commonly means:
    a. it has never been used
    b. it is damaged
    c. it is a good hose for reuse
    d. rolling style means nothing

15. The text suggests that upon reaching the fire area, approximately how many feet of working line should be left over?
    a. 50 ft
    b. 100 ft
    c. 200 ft
    d. none needed

16. A preconnected load means that the hose is:
    a. already attached to a pump discharge
    b. pre-coupled to an intake relief valve
    c. assembled for a standpipe operation
    d. made ready for easy deployment at a hydrant

17. When advancing an uncharged hose line upstairs, it is best to:
    a. keep the hose close to the inside edge or rail of the stair
    b. lay the hose against the outside edge or rail of the stair
    c. wrap the hose around hand rails to help keep the hose from sliding downward
    d. run excess hose below the level of the fire to help protect it

18. Portable deluge sets should not operate at an angle below:
    a. 75 degrees
    b. 50 degrees
    c. 40 degrees
    d. 25 degrees

## Fill-in-the-Blanks

    a. brush
    b. detergent
    c. internal
    d. gasket
    e. sharp
    f. higbee
    g. roller
    h. grit
    i. external
    j. rough

A number of things can be done to keep hose in service for a very long time. For instance, one should avoid laying hose over (1) _____ or (2) _____ corners, and a hose (3) _____ should be used if available. Hose should also be washed with a(n) (4) _____ and water in order to remove (5) _____ from the jacket. If the hose has been exposed to chemicals, then a(n) (6) _____ can be used to assist in the cleaning.

When coupling hose, one will notice that there are both (7) _____ and (8) _____ threads. A(n) (9) _____ will be found on the threads to assist the firefighter. Firefighters are taught to feel the inside of the female coupling for a(n) (10) _____ before making the connection.

11. Fill in the name of the hose loads on the apparatus in Figure 10-2.

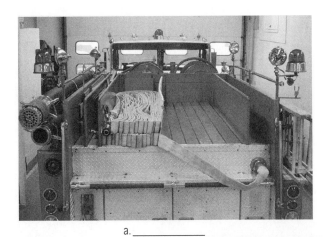

a._____

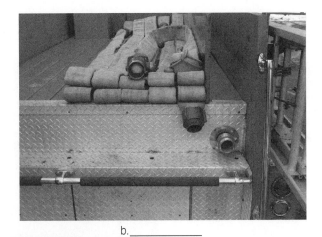

b._____

c._____

**Figure 10-2**   Name the hose loads above.

12. Arrange the photo sequence in Figure 10-3 and describe the operation.

(_____)_____

(_____)_____

(_____)_____

a.

b.

c.

**Figure 10-3** Arrange and describe this sequence.

13. Describe the operation in Figure 10-4.

_____

_____

_____

a.

b.

c.

d.

**Figure 10-4**  Describe this operation.

14. What is the hose lay in Figure 10-5 called and where is it used most often?

_____

_____

_____

_____

**Figure 10-5** Describe this operation.

# CHAPTER 11

# NOZZLES, FIRE STREAMS, AND FOAM

# QUESTIONS

## Matching

Match the correct term with the definitions provided.

   a. nozzle reach

   b. fire stream

   c. nozzles

   d. fog nozzle

   e. nozzle reaction

1. A _____ is the flow of water or other agents from the nozzle toward a target.

2. Appliances that apply water or extinguishing agents are called _____ .

3. The distance that water will travel after leaving a nozzle is called _____ .

4. _____ is the force of nature that makes the nozzle move in the opposite direction of the water flow.

5. The _____ delivers either a fixed spray pattern or a variable pattern to the target.

6. Match the nozzles listed below with the labels in Figure 11-1.

   a. _____

   b. _____

   c. _____

   d. _____

   rotating with playpipe

   built-in lever with pistol grip

   break-apart playpipe

   built-in lever

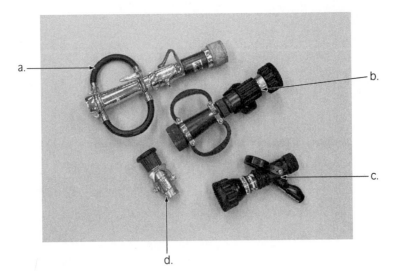

**Figure 11-1**  Match the terms to the letters.

7. Match the foam application styles listed below to the photos in Figure 11-2.

    a. _____

    b. _____

    c. _____

raindown technique

bank-in technique

bank-back technique

a. _____

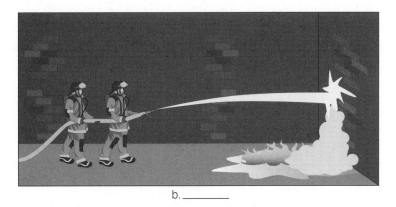

b. _____

c. _____

**Figure 11-2** Match the foam application style to the drawing.

8. Write in the terms as they apply to Figure 11-3 describing nozzle reaction.

Water flow

Solid stream of water

Restriction increases pressure

Pressure

Velocity

Nozzle reaction

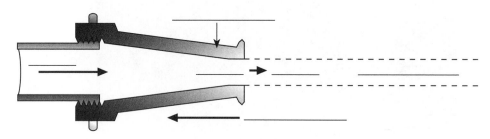

**Figure 11-3** Match the terms.

# True or False

1. In today's fire service, nozzles are seldom used at their maximum performance capabilities.
   a. true    b. false

2. All of today's nozzles can adjust the flow either manually or automatically.
   a. true    b. false

3. Maximum horizontal reach is between a 65- to 70-degree angle.
   a. true    b. false

4. Solid stream handlines have the ability to penetrate through the fire's heat without absorbing that heat before reaching the target.
   a. true    b. false

5. It is possible to use the fog nozzle to assist with horizontal ventilation.
   a. true    b. false

# Multiple Choice

1. Name the two kinds of nozzles most commonly used in today's fire service. (choose two)
   a. interior
   b. deck gun
   c. fog
   d. straight stream

2. Which of the following factors most often affects the reach of a stream? (choose two)

a. gravity

b. water clarity

c. stream shape

d. mineral content of water

3. Which nozzle delivers an unbroken or solid stream of water at the tip?

a. solid stream

b. fog

c. foam

d. chemical

4. What is the advantage of the smooth bore nozzle over other nozzles?

a. less parts to maintain

b. tuberculation

c. venturi effect

d. mobility

5. Which of the following streams can produce more steam when applied properly?

a. fog

b. smooth

c. master

d. combination

6. Which of the following streams can be used for moving large volumes of air?

a. fog

b. smooth

c. master

d. combination

7. The Bresnan distributor is most commonly used for what type of firefighting?

a. high-rise

b. rubbish

c. cellar

d. aircraft

8. Piercing nozzles were originally designed for what type of firefighting?

a. high-rise

b. rubbish

c. cellar

d. aircraft

9. What type of nozzle is used for flowing large volumes of water up to and exceeding 1,000 gpm?

a. master stream

b. large reach

c. high pressure

d. big gun

10. When figuring hydraulics, it is common practice to express pressure in which of the following ways?
    a. psi
    b. inches
    c. feet
    d. gpm

11. Pressure readings on the pump panel are taken from which of the following?
    a. vacuum pressure
    b. atmospheric pressure
    c. gauge pressure
    d. none of the above

12. What term is used for the loss of energy due to turbulence or rubbing of the moving water through the hose?
    a. pressure peak
    b. friction loss
    c. relative pressure
    d. static pressure

13. Name two natural factors that affect fire streams. (choose two)
    a. pump failure
    b. gravity
    c. nozzle reaction
    d. wind

14. An aggregate of gas-filled bubbles is called:
    a. friction
    b. pressure
    c. foam
    d. fog

15. What term is used when discussing the amount of foam or foam solution needed to extinguish a fire?
    a. pattern
    b. activity
    c. volume pressure
    d. application rate

16. A detergent-based foam that reduces water surface tension is used on what type of fire?
    a. Class A
    b. Class B
    c. flammable gases
    d. gasoline

17. Class B foam is most often used on fuels such as:
    a. wildland fuels
    b. common combustibles
    c. water soluble fuels
    d. gases

18. A common foam proportioner using the venturi principle is called a(n):
    a. distributor
    b. eductor
    c. fog
    d. expansion

19. Compressed air foam systems typically use what type of foam?
    a. alcohol
    b. protein
    c. Class A
    d. Class B

20. What type of nozzles are shown in Figure 11-4?
    a. fog
    b. foam
    c. straight stream
    d. master stream

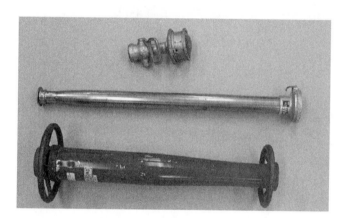

**Figure 11-4**  Nozzles.

21. The fire stream from a solid stream handline can reach a maximum of approximately:
    a. 30 feet
    b. 50 feet
    c. 70 feet
    d. 100 feet

22. All of the following can effect the friction loss within a hoseline *except:*
    a. hose diameter
    b. nozzle type
    c. flow quantity
    d. hose length

23. Alcohol, lacquer thinner, and acetone can break down ordinary Class B foam because they are miscible in water. These liquids are classified as:

    a. polar solvents

    b. hydrocarbons

    c. polymerics

    d. A-type fuels

# Fill-in-the-Blanks

    a. indirect

    b. nozzle flow

    c. fog

    d. direct

    e. nozzle reach

When selecting a nozzle for an attack it is important that (1) _____ and (2) _____ be considered in determining effective operation. This is especially true when considering the type of attack. If the attack will be flowing water directly on to the seat of the fire, it is a(n) (3) _____ attack, and if it is going to be where a(n) (4) _____ is used for steam in a closed environment, it is a(n) (5) _____ attack.

# CHAPTER 12

# PROTECTIVE SYSTEMS

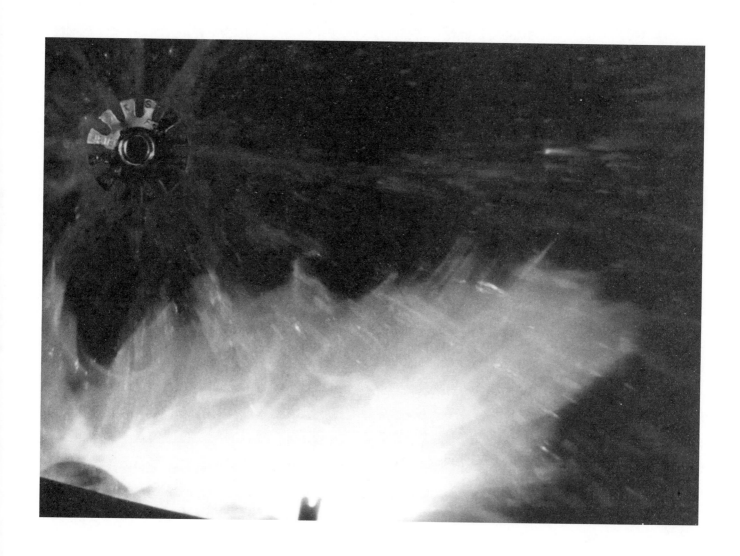

# QUESTIONS

## Matching

Match the correct term with the definitions provided.

    a.  ionization
    b.  standpipes
    c.  photoelectric
    d.  detection
    e.  heat detectors
    f.  red
    g.  uncolored
    h.  green
    i.  blue
    j.  white

1.  When discussing fixed suppression systems, the two main systems are sprinklers and
    _____ .

2.  The _____ system will discover the fire or other problem and notify people of the
    situation.

3.  _____ operate(s) by detecting the heat of a fire and judging its rate of rise.

4.  Detectors that use a radioactive element are _____ detectors.

5.  The detector that utilizes light is called a(n) _____ detector.

6.  Using color codes for sprinkler heads has been around for some time. _____ denote(s)
    an ordinary head with a temperature rating of 135° to 170°F.

7.  An intermediate head (175° to 225°F) is colored _____ .

8.  A high head (250° to 300°F) is colored _____ .

9.  An extra high head (325° to 375°F) is colored _____ .

10. A very high head (400° to 475°F) is colored _____ .

11. In Figure 12-1, two sprinkler heads are portrayed. Which head in the figure is pendent, a or b? (circle one)

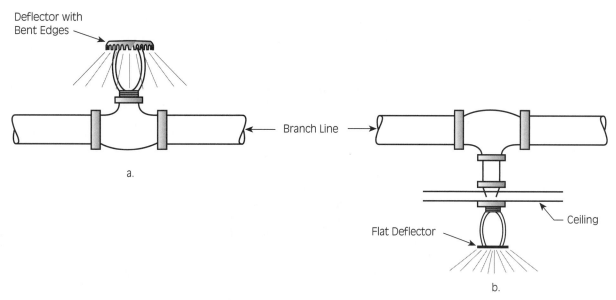

**Figure 12-1** Which head in this figure is a pendent head?

12. Match the wet pipe sprinkler system terms to the system in Figure 12-2.

Automatic sprinklers

Water motor alarm (water gong)

Check valve

Fire department connection

Main drain connection

OS&Y valve

Inspector's test connection

Riser Alarm valve

Riser

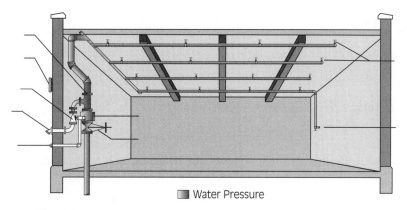

**Figure 12-2** Wet pipe sprinkler system.

13. Identify the parts of the sprinkler piping diagram in Figure 12-3 using the following terms.
    1. Branch line
    2. Riser
    3. Cross main
    4. Feed main

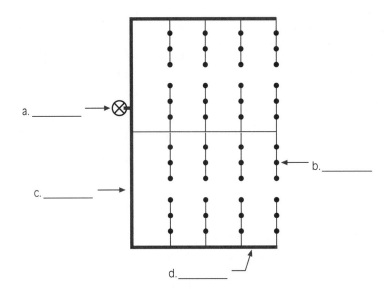

**Figure 12-3** Sprinkler piping diagram.

# True or False

1. Detection systems may not only detect the presence of a fire but may also activate a suppression system.
   a. true     b. false

2. Builders have gone to great lengths to convince the fire service of the need for fire prevention systems and codes.
   a. true     b. false

3. Proper signage at pull stations can help to teach people about the need to call the fire department in case of emergency.
   a. true     b. false

4. Heat detectors are considered the best type of safety device because of their speedy reaction to fire.
   a. true     b. false

5. Heat detectors are relatively inexpensive and have a low rate of false alarms.
   a. true     b. false

6. Smoke and toxic gases are the leading causes of death for people in residential occupancies.
   a. true     b. false

7. Gas detectors are somewhat unique, since they can be either permanently mounted or portable.
   a. true     b. false

8. Sprinkler systems are between 90 percent and 100 percent effective in stopping the spread of fire.

    a. true      b. false

9. The primary purpose of residential sprinklers is to prevent destructive fire and not necessarily to save lives, since that job is for the smoke detector.

    a. true      b. false

10. According to the text, the wet pipe sprinkler system allows the quickest response when the head is opened.

    a. true      b. false

# Multiple Choice

1. Which two of the following devices are said to allow water or other extinguishing agents to be used quite effectively? (choose two)

    a. PPE

    b. standpipes

    c. pumps

    d. sprinklers

2. Detection systems discover fire quickly and are designed to give warning to all but which of the following?

    a. building occupants

    b. alarm companies

    c. fire departments

    d. building landlords

3. What type of fire alarm system requires a person to pull a lever or push a button?

    a. fixed

    b. manual

    c. wet

    d. dry

4. Street box systems are which type of alarm system?

    a. fixed

    b. manual

    c. wet

    d. dry

5. There are two problems that are typical with manual systems. What are they? (choose two)

    a. they are hard to maintain

    b. a person has to activate them

    c. they are usually only local alarms

    d. they break down continually

6. Name the two types of residential smoke detectors. (choose two)
   a. manual
   b. photoelectric
   c. ionization
   d. rate of rise

7. Which of the following is *not* really a type of flame detection device?
   a. light obscuration
   b. ultraviolet
   c. infrared
   d. combined UV and IR

8. Statistics show that most of the structure fires with the most fire damage occur in which type of occupancy?
   a. explosive
   b. retail
   c. manufacturing
   d. residential

9. Some sprinklers are upright and some are:
   a. obscure
   b. retracted
   c. pendent
   d. closed

10. The orifice or size of the water opening in the typical sprinkler head varies from:
    a. ¼ to ¾ inch in diameter
    b. ½ to 1 inch in diameter
    c. ¾ to 1¼ inches in diameter
    d. 1 to 1½ inches in diameter

11. Temperature ratings of sprinkler heads range from a low of:
    a. 50°F
    b. 100°F
    c. 135°F
    d. 175°F

12. The "ordinary" rated head is the most common, and it averages what temperature rating?
    a. 75°F
    b. 165°F
    c. 210°F
    d. 312°F

13. Which of the following is *not* listed in the text as one of the three types of fusible elements on sprinkler heads?
    a. fusible link
    b. bulb
    c. chemical pellet
    d. deluge

14. Which of the following types of sprinkler systems is considered the simplest in design and operation?
    a. wet pipe
    b. dry pipe
    c. combo type
    d. chemical

15. The main control valve used to shut off the supply of water to the entire system is usually called the:
    a. clapper valve
    b. test valve
    c. drain valve
    d. OS&Y

16. One would find the dry pipe system usually in which type of building?
    a. outside
    b. common combustible
    c. unheated
    d. warehouse

17. Which system is designed so that all of the sprinkler heads flow at the same time?
    a. wet pipe
    b. quick action
    c. deluge
    d. flood

18. Which type of sprinkler system would one most often find in computer rooms and historical document storage areas?
    a. deluge
    b. preaction
    c. wet pipe
    d. quick action

19. If the water source to the sprinkler system is lacking in ability to supply large quantities to a large system, which of the following is most often added to improve the flow?
    a. extra hydrants
    b. OS&Y valves
    c. fire pump
    d. security guards

20. A type of vertical pipeline built to assist firefighters with hoselines in a high-rise building is called:
    a. wall pipe
    b. riser
    c. check valve
    d. standpipe

21. An intermittent alarm from a carbon monoxide detector could indicate:
    a. that the unit is not functioning properly
    b. that life-threatening levels of CO are present
    c. an early warning that low-level CO is present
    d. that high temperature CO is present

22. The activation of a deluge system is triggered by:
    a. the activation of several detectors
    b. the fusing of several sprinkler heads
    c. pushing the dead-man switch
    d. firefighter intervention only

23. Firefighters should silence loud audible alarms:
    a. upon arrival when investigating alarms
    b. only after meeting with the building representative
    c. when they interfere with hearing radio transmissions
    d. after investigating and finding no fire in the building

## Fill-in-the-Blanks

a. photoelectric
b. smoke detector
c. flame detector
d. ionization
e. gas detector

Fire departments are finding that their job is being made safer with the advent of detectors, giving occupants quicker warning of fire. The (1) _____ is one such device. It comes in two types: (2) _____ , which senses smoke using radioactive elements, and (3) _____ , which detects smoke with light refraction. Another detector used in more industrial settings is the (4) _____ . These are commonly hooked up to "smell" the air. The detector that is configured to sense the ultraviolet and infrared waves of flame is termed the (5) _____ .

6. Name the three types of sprinklers in Figure 12-4.

a._____

b._____

c._____

**Figure 12-4** Types of sprinklers.

7. Which of the OS&Y valves in Figure 12-5 is in the open position?
   a or b (circle one)

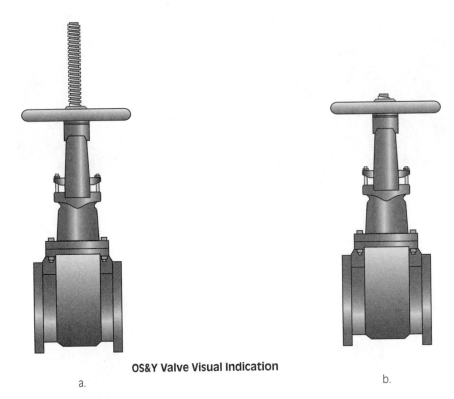

**OS&Y Valve Visual Indication**

a.                                              b.

**Figure 12-5**   OS&Y valve visual indication.

8. Which of the standpipe systems in Figure 12-6 is considered a Class I system?
   a or b (circle one)

a.                                              b.

**Figure 12-6**   Standpipe systems.

# BUILDING CONSTRUCTION

# QUESTIONS

## Matching

Match the correct term with the definitions provided.

     a. Type V

     b. Type II

     c. Type I

     d. Type III

     e. Type IV

1. _____ construction shall be that type in which the structural members are of approved noncombustible or limited combustible materials.

2. The structure type known as _____ shall consist of structural members that are approved noncombustible or limited combustible materials, and the interior can be wood.

3. _____ construction shall be that type in which exterior walls can be constructed entirely or partially of wood or other approved combustible material.

4. Heavy timber construction is classed _____ construction.

5. _____ construction is much like Type I; however, it is termed *noncombustible construction*.

6. Loads are applied to a structural member as tension, shear, and compression forces. Identify these forces in Figure 13-1.

    a. _____    b. _____    c. _____

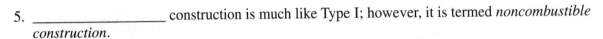

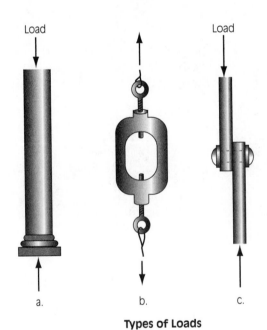

**Types of Loads**

**Figure 13-1**   Identify the types of loads as shown in a, b, and c.

7. Match the common roof framing styles to the drawings in Figure 13-2.

| | |
|---|---|
| hip | intersecting |
| shed | gambrel |
| gable | butterfly |
| mansard | |

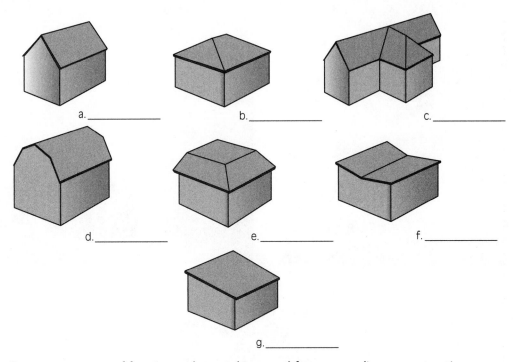

a._____    b._____    c._____

d._____    e._____    f._____

g._____

**Figure 13-2**   Some common roof framing styles used in wood frame or ordinary construction.

8. Which of the structures in Figure 13-3 is considered to be Type III construction?

   Circle one:   a,   b,   or   c

a.

b.

c.

**Figure 13-3**   Common structural types.

# True or False

1. Firefighters must understand that fire-resistive ratings are done in a laboratory and therefore are very accurate in the field.

   a. true     b. false

2. An important item for firefighters to remember is that building owners may modify their structures without engineering or building department knowledge.

   a. true     b. false

FFD 3. Steel is somewhat unique because it may expand with heat, causing structural instability.

   a. true     b. false

FFD 4. Steel has been known to actually stop the spread of fire because of its ability to not conduct heat to other building areas.

   a. true     b. false

5. One must be aware of buildings under construction since they are made up of parts possibly not connected until finished. Collapse potential may be greater.

   a. true     b. false

6. Masonry has very little lateral stability, and in many cases the roof or floor structure of a building holds the walls in place.

   a. true     b. false

7. The most common building material in use today for building construction is probably still steel.

   a. true     b. false

8. Trusses have been a danger to firefighters for some time; however, from an engineering standpoint, the truss is a sound structural element.

   a. true     b. false

FFD 9. The combination of loading and a truss weakened by fire will cause a very quick collapse during firefighting operations.

   a. true     b. false

# Multiple Choice

1. It is said that fire-resistive ratings, means of egress, and which of the following are essential building code provisions allowing life safety?

   a. stairwell support

   b. elevator controls

   c. construction evaluation

   d. occupancy classifications

2. The building industry claims that fire-resistive ratings for the following type of occupancy run from three to four hours.

   a. wood frame

   b. truss supported

   c. fire resistive

   d. balloon construction

3. Wood framing is done in a few different ways, with which two of the following considered the most common? (choose two)

 a. platform

 b. prefabricated

 c. fire resistive

 d. balloon

4. Which of the following framing types provides for fire-stopping at each level or story?

 a. platform

 b. prefabricated

 c. fire resistive

 d. balloon

5. Beams, girders, large trusses, and lintels are made of both wood and which of the following?

 a. masonry

 b. steel

 c. rock

 d. glass

6. A temperature of 1,100°F will greatly affect steel. It is said that steel exposed to this temperature can lose up to how much of its strength?

 a. 10 percent

 b. 25 percent

 c. 50 percent

 d. temperature has no effect

7. At about 1,000°F, steel will begin to expand. It can expand as much as how far for each 10 feet of total length?

 a. 1 foot

 b. 8 inches

 c. 5 inches

 d. 1 inch

8. The building industry commonly refers to brick, concrete block, and stone as:

 a. mortar

 b. veneer

 c. masonry

 d. heavy stuff

9. What is the weakest point of a block wall?

 a. mortar

 b. corner

 c. bottom

 d. top

10. Which type of construction would be most likely to use masonry load-bearing walls?
    a. wood frame
    b. ordinary construction
    c. special hazard
    d. they are never allowed

11. Occupancies are commonly broken down into five types, with which of the following not being one of those types?
    a. business
    b. industrial
    c. aircraft
    d. residential

12. Some signs of building collapse potential are listed below, with which one having the least potential?
    a. cracks in walls
    b. bulging walls
    c. deteriorated condition
    d. fire-stops removed

13. The absolute minimum distance one must work from a building, in case of its collapse, is said to be at least:
    a. 1½ times the height of the building
    b. 3 times the height of the building
    c. 4 times the height of the building
    d. there is no minimum

14. A new composite building material is called "SIP". SIP stands for:
    a. Structural Insulated Panels
    b. Strand-board Integrated Products
    c. Self-Interlocking Panels
    d. Styrofoam Infused Products

15. A parapet is best described as a(n):
    a. unsupported cantilevered beam
    b. free-standing masonry wall past the top of a roof
    c. distinctive roof style with intersecting pitches
    d. void space capable of extending fire spread

16. Lightweight wood products are typically fashioned from:
    a. wood chips and glue
    b. genetically altered trees
    c. carbon fibers and ore
    d. composite-reinforced polystyrene

## Fill-in-the-Blanks

a. fire load

b. loading

c. impact load

d. dead load

e. concentrated load

As a firefighter one must know how a building works, how it stands, and how it holds its weight. The term associated with the weight of a building is called (1) _____ . For instance, a load applied to a very small area in a building may be called its (2) _____, and the load of the entire building and the building materials is its (3) _____ . As fire units arrive on scene and begin the attack, they will most commonly be concerned with the (4) _____ because the contents of the building are where most fires originate. As master streams are applied, the (5) _____ becomes a consideration since the danger of collapse must be considered.

6. Identify the kind of roof construction being put up in Figure 13-4.

_____

**Figure 13-4**   A typical Type V wood frame structure during construction.

7. What are the pieces called that are meant to hold this parallel chord truss together in Figure 13-5?

_____

**Figure 13-5**   A typical parallel chord truss.

# QUESTIONS

## Matching

Match the correct term with the definitions provided.

     a. hooks

     b. fly

     c. heel

     d. beam

     e. halyard

1. The _____ is the part of a ladder that runs along each side from bottom to top.

2. The bottom of the ladder where it rests on the ground is most often called the _____ .

3. A _____ runs through a pulley giving the firefighter the ability to raise the extension ladder.

4. A set of retractable _____ is mounted on the end of the roof ladder so it can be placed safely on a steep surface without slipping down.

5. An extension ladder has multiple sections, with the sections not at the bottom most often called the _____ sections.

6. Match the following ladder raising terms to the spaces in Figure 14-1.

   rotate

   raise/lower

   extend/retract

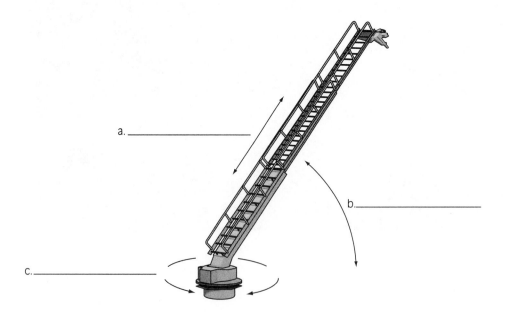

**Figure 14-1** Ladder positioning terminology.

7. Match the ladder carry term to the letter under the correct photo in Figure 14-2.

   shoulder

   suitcase

   flat

   a. _____ b. _____ c. _____

a.

b.

c.

**Figure 14-2** Name the ladder carries.

# True or False

1. Because ladders have grown in height, the supporting beams have been manufactured most recently with truss type units for greater strength.

   a. true      b. false

2. The wood beams have been replaced with steel and aluminum on the longer ladders used on truck companies.

   a. true      b. false

3. Every fire department in the country uses standard ladder terminology, so any firefighter can learn everyone's terms easily.

   a. true      b. false

4. Technically speaking, tower ladders and articulated beam ladders are not ladders.
   a. true      b. false

5. Aerial ladders have been manufactured to raise as high as 144 feet.
   a. true      b. false

6. The main advantage of the aerial ladder is that it can hold many people at the same time and can work as an observation vantage point as well as a work platform.
   a. true      b. false

7. The advantage of the snorkel ladder is that the basket can be lowered behind an obstruction, while a tower ladder cannot.
   a. true      b. false

8. The Pompier ladder has been aproved by the NFPA as a standard ladder.
   a. true      b. false

9. It is recommended that firefighters should remove their gloves when operating ladders as gloves do not grip well and may allow the ladder to slip during raising and lowering.
   a. true      b. false

10. All ladders are electrical conductors no matter what their construction type may be.
    a. true      b. false

11. The tip firefighter controls the operation and gives the commands.
    a. true      b. false

12. The ideal safe climbing angle is 70 degrees.
    a. true      b. false

13. When ascending a ladder, three limbs should remain in contact with the ladder at all times.
    a. true      b. false

## Multiple Choice

1. Ladders were originally constructed of which of the following materials in the early fire service?
   a. steel
   b. aluminum
   c. wood
   d. hard rubber

2. Which of the following terms is used for turning the ladder from left to right?
   a. raise
   b. rotate
   c. extend
   d. retract

3. Which of the following terms is used for decreasing the reach of the ladder by nesting the fly sections?

   a. raise

   b. rotate

   c. extend

   d. retract

4. The extension ladder's movement, when raising, is achieved through which of the following parts? (choose two)

   a. beam

   b. cables

   c. rungs

   d. pulleys

5. Which of the following is not a common type of ground ladder?

   a. tower

   b. attic

   c. straight

   d. extension

6. When cleaning most ladders, the best thing to use is: (choose two)

   a. detergent

   b. steel wool

   c. diesel fuel

   d. scrub brush

7. According to the text, which of the following poor practices accounts for the majority of ladder-related injuries?

   a. not locking in

   b. tying off

   c. overloading

   d. not footing

8. What is the most common use of a ladder?

   a. bridging

   b. diking

   c. rescue

   d. access

9. The rule of thumb for using ladders as bridges is that for each 10-foot span the support points at each end must extend:

   a. 5 feet

   b. 10 feet

   c. 15 feet

   d. 20 feet

10. When selecting a ladder, the first consideration should be:
    a. location of use
    b. overhead obstructions
    c. ground surface
    d. number of firefighters

11. If the ladder will be used for access or escape, the tip should extend how far above the sill?
    a. five rungs
    b. three rungs
    c. two rungs
    d. no rungs

12. When placed for access to a roof, the text recommends that the tip should extend how far above the roofline?
    a. five rungs
    b. three rungs
    c. two rungs
    d. no rungs

13. The standard formula for determining the proper distance the base of the ladder should be from the building is:
    a. one-half the height of the building
    b. one-fourth the height of the ladder
    c. 10 feet
    d. 5 feet

14. Ladder carries differ for the type of ladder. Which of the following types of carries are recommended by the text? (choose two)
    a. shoulder
    b. drag
    c. suitcase
    d. knot

15. Name the two methods suggested by the text for raising ground ladders. (choose two)
    a. throw
    b. thrust
    c. beam
    d. rung

16. The term for the ladder raise in Figure 14-3 is:

    a. quick raise

    b. attic raise

    c. rafter raise

    d. bangor raise

**Figure 14-3** Name this ladder raise.

17. After an extension ladder is raised, and before it is lowered into the building, the firefighter at the butt of the ladder should:

    a. wait for a fourth firefighter to command the "lower ladder" function

    b. tie off the halyard for stafety

    c. track positions with the tip firefighter

    d. begin the climb

# Fill-in-the-Blanks

a. turntable        b. fly

c. degrees        d. pistons

e. bed

Aerial ladders are designed so that various sections slide out from each other. The construction of each (1) _____ section is designed to overlap the section below, with the lowest section being the (2) _____ ladder. This lower section is attached to the apparatus with a combination of pins and (3) _____ to allow for movement. This rotating platform is called the (4) _____ , and it may rotate as far as 360 (5) _____ .

6. Fill in the ladder terms in Figure 14-4 using the following nomenclature.

| | |
|---|---|
| beam | tip |
| rungs | gusset plates |
| butts | rails |
| fly section | pulley |
| guides | halyard |
| ladder locks | bed section |

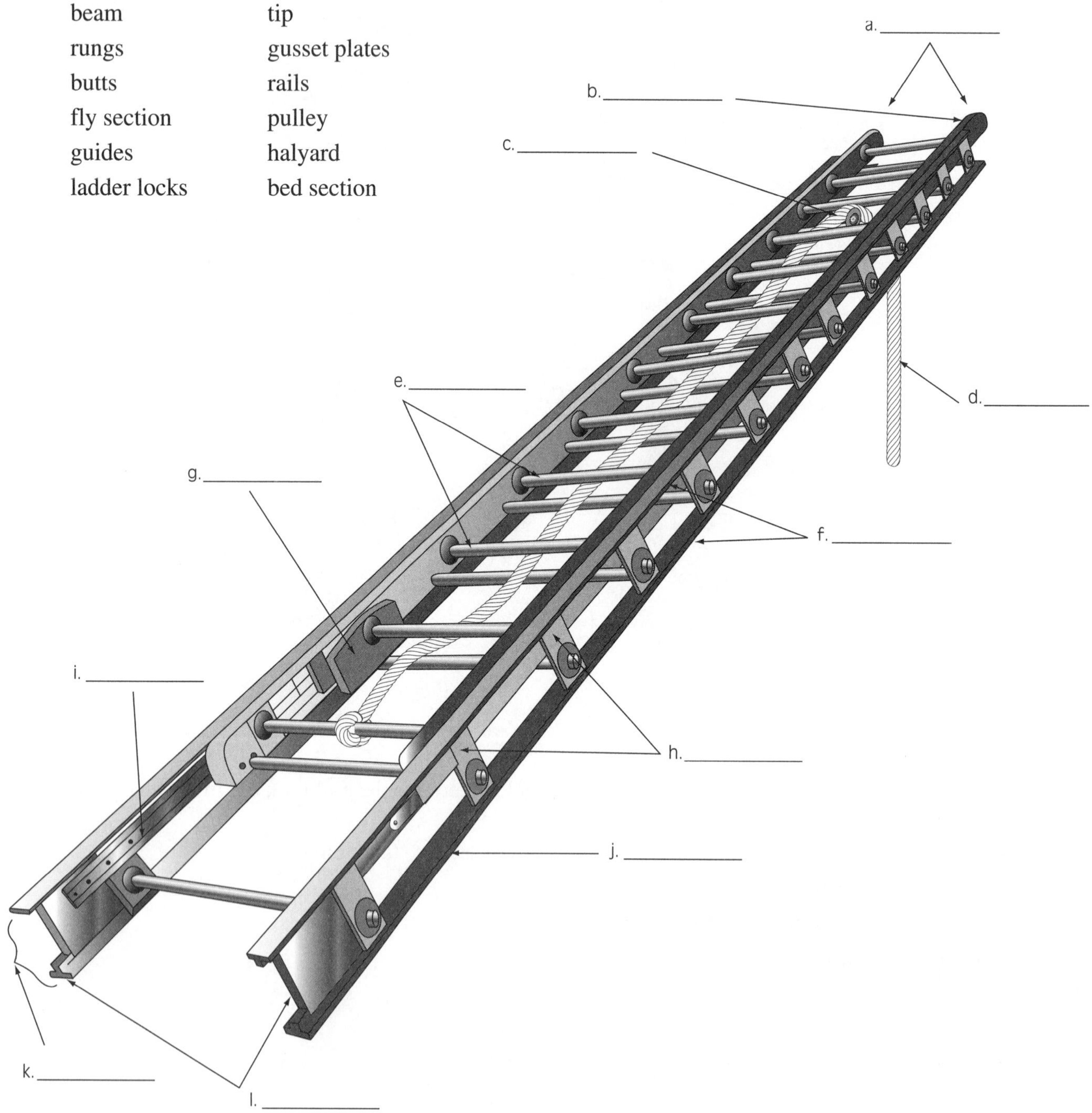

**Figure 14-4**   Fill in the blanks using terms in question 6.

7. The ladder used in Figure 14-5 is very common in confined space rescue. What is it called?

_____

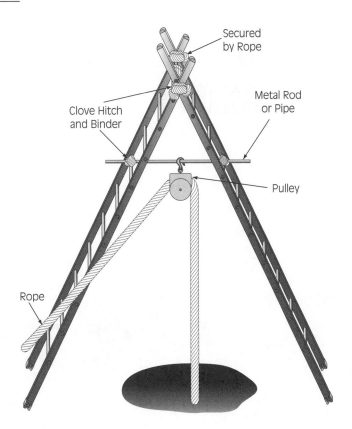

**Figure 14-5** Name this ladder operation.

8. The operation the firefighter in Figure 14-6 is performing is called _____.

**Figure 14-6** Name this operation.

# ROPES
# AND KNOTS

# QUESTIONS

## Matching

Match the correct term with the definition provided.

    a. static

    b. laid

    c. 1

    d. dynamic

    e. polyester

1. Manila rope is available in different types, with Type _____ being the highest quality.

2. The rope fiber that withstands the highest temperatures among the ropes used in the fire service is _____ .

3. _____ ropes have very little elongation at normal safe working loads.

4. Rope with a high degree of elongation at normal safe working loads is termed _____ rope.

5. The most common type of rope construction for natural fiber rope is termed _____ .

6. Figure 15-1 identifies the elements of a common knot. Match these elements from left to right in the figure using the following terms:

    bight

    round turn

    loop

    a._____      b._____      c._____

**Figure 15-1**   Identify the common elements of a knot using the terms listed in question 6.

7. Match the names of the knots to the knots shown in Figure 15-2.

    double becket bend         figure eight         bowline         half hitch

    a. _____ b. _____ c. _____ d. _____

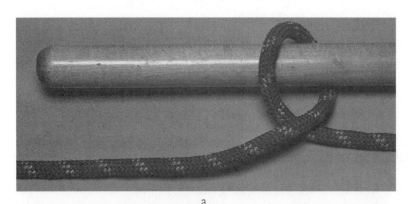

a.

b.

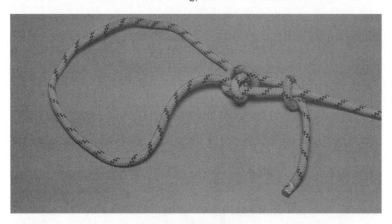

c.

d.

**Figure 15-2** Match the names of these knots.

# True or False

1. Natural material rope is quite resistant to mildew, whereas synthetic rope is not.
   a. true        b. false

2. Rope manufactured from natural materials should be used for utility line.
   a. true        b. false

3. Soaking new manila rope in water will increase its strength as it dries.
   a. true        b. false

4. The fire service of today uses more synthetic rope than natural rope.
   a. true        b. false

5. Two positive features of nylon rope are that it has a high melting point and excellent abrasion resistance.
   a. true        b. false

6. NFPA lists only two types of rope that can be used as both utility and life safety line at the same time.
   a. true        b. false

7. The bowline does not hold as well as with the newer synthetic ropes.
   a. true        b. false

8. If a life safety rope is damaged or suspect, it should be immediately removed from service.
   a. true        b. false

9. When washing rope in a washing machine, a top-loading machine is better because the rope sits well without being twisted.
   a. true        b. false

10. One of the primary uses of rope on the fire scene is hoisting tools.
    a. true        b. false

11. The working end is the end of the rope utilized to tie the knot.
    a. true        b. false

12. The standing part of the rope is between the working end and the running end.
    a. true        b. false

13. The running end is used to tie the knot.
    a. true        b. false

# Multiple Choice

1. Which NFPA document set the minimum standards for rope used by firefighters during the performance of their duties?
   a. 1500
   b. 1900
   c. 1983
   d. 2002

2. Which of the following materials is not common to natural rope?

   a. nylon

   b. sisal

   c. cotton

   d. manila

3. The best way to tell Type 1 manila rope is by which of the following identifiers?

   a. braid type

   b. inner line

   c. colored string

   d. spliced ends

4. Ropes made of cotton fiber are lower in strength than manila by what percent?

   a. 10 percent

   b. 25 percent

   c. 50 percent

   d. 75 percent

5. The primary materials utilized in synthetic rope are all but which of the following?

   a. nylon

   b. polyester

   c. polyvinyl

   d. polyethylene

6. Nylon line is used most often in which of the following applications?

   a. fishing line

   b. stockings

   c. hose

   d. fiber optics

7. The primary use of polypropylene rope is:

   a. hoisting tools aloft

   b. high angle rescue

   c. HAZMAT rescue

   d. water rescue

8. Some of the modern rope construction types are: (choose two)

   a. braided

   b. lined

   c. kernmantle

   d. weaved

9. Making sure that all the parts of the knot are lying in the proper orientation is called:

   a. tying

   b. sorting

   c. chafing

   d. dressing

10. When tying webbing, only one type of knot is common. Which of the following is that knot?

    a. figure eight

    b. bowline

    c. water

    d. cinch

## Fill-in-the-Blanks

    a. running end

    b. bight

    c. working end

    d. setting

    e. standing part

A length of rope is divided into three sections. The (1) _____ is the end of the rope utilized to tie the knot, while the (2) _____ is the rest of the rope not used in tying. The (3) _____ is the end that is not rigged or tied off. When the rope forms a U-turn utilizing a double section and does not cross itself, it is termed a (4) _____ . When the knot is tied and dressed, the last thing to do is (5) _____ the knot by pulling it snug.

6. Figure 15-3 shows a rope with the working end, the running end, and the starting part. Put the correct term into the space provided.

**Figure 15-3** Put the terms in the spaces provided.

7. Figure 15-4 shows a knot that has been properly tied, _____ and _____.

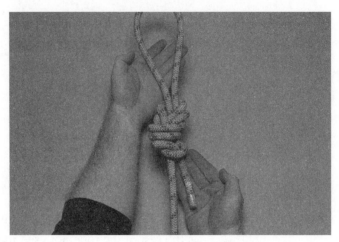

**Figure 15-4**   A properly tied knot.

# RESCUE PROCEDURES

# QUESTIONS

## Matching

Match the correct term with the definition provided.

    a. right hand

    b. backboards

    c. tunnel vision

    d. two in/two out

    e. benchmarks

1. It is generally referred to as _____ when firefighters begin to focus on one item, losing sight of other details that may affect their safety.

2. OSHA created a rule, called the _____ rule, where firefighter safety is of paramount importance in interior firefighting conditions.

3. A search term used when using walls as bearing points is termed the _____ method.

4. The primary and secondary search operations are typically called tactical _____ when utilized by the incident commander.

5. When rescuing victims, _____ are designed to provide the maximum in spinal immobilization.

6. Figure 16-1 portrays three types of structural collapse. Match the types to the drawings in the figure.

    a. _____

    b. _____

    c. _____

    V-type

    pancake

    lean-to

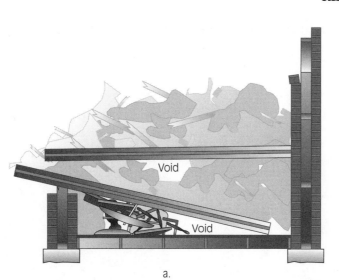

a.

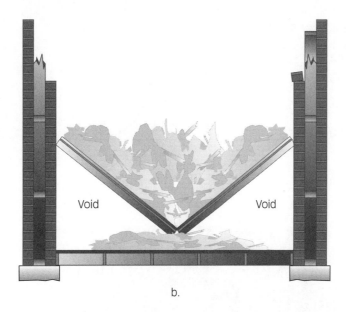

b.

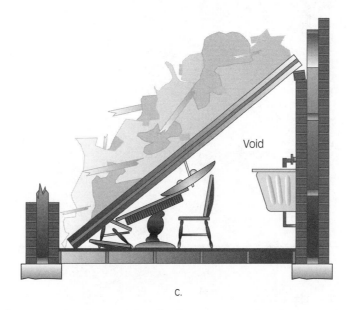

c.

**Figure 16-1** Three types of structural collapse. Identify these types using the terms in question 6.

7. Figure 16-2 shows three different ways of carrying patients. Match the terms below to the carries in the figure.

firefighter's carry

seat carry

extremity carry

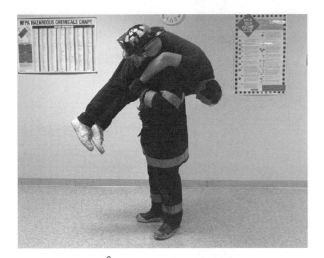

a. _____

b. _____

c. _____

**Figure 16-2**  Name these carries using the terms provided.

# True or False

1. Tunnel vision is not much of a concern in rescue situations because of today's advanced technology.
   a. true    b. false

2. Whenever interior firefighting is to take place, a minimum of two firefighters must be standing by for possible rescue of the team members inside.
   a. true    b. false

3. Lifeline rope can be used as a guideline because the basic premise of life safety is the same.
   a. true    b. false

4. During the primary search of a structure, the firefighting team must stay behind the fire attack team for safety.
   a. true    b. false

5. When rescuing a downed firefighter whose SCBA is not working, it is a good idea to disconnect the person's air hose from the regulator and place it into the person's turnout coat.
   a. true    b. false

6. Most of the firefighter rescue drags are designed to provide a simple form of spinal immobilization for the patient.
   a. true    b. false

7. It is estimated that motor vehicle accidents are probably the most common rescue situations that today's firefighters will be involved in.
   a. true    b. false

8. Power tools require hydraulic fluid, which makes their maintenance quite easy since the same fluid can be used in a multitude of tools on the common rescue squad.
   a. true    b. false

9. Eye protection is no longer mandated when using power hydraulic tools due to the heavy construction of today's tools.
   a. true    b. false

10. When using rescue air bags, it is important that they rest on a solid base.
    a. true    b. false

11. It is a practice in many areas that the first rescuer in a water rescue may jump in without a PFD but that all follow-up rescuers must have them on upon arrival.
    a. true    b. false

12. The cave-in victim usually dies of asphyxiation even though the victim's head may not have been buried.
    a. true    b. false

13. It is common knowledge that the first few seconds of rescue are the most important in cave-in rescues. Therefore, the first responder on scene will go directly into the hole and make a quick attempt to free the victim, and if unsuccessful will pull back to set up the operation for an extended rescue.
    a. true    b. false

14. It is estimated that up to one-half of the victims of confined space rescues are those who attempted to rescue a primary victim.

    a. true    b. false

15. Two wires are lying in the street. The firefighter can tell which one is energized by the sparks or current coming from one of them and not the other.

    a. true    b. false

16. The "independent service" switch in an elevator will cause the car to be controlled only from the ground floor.

    a. true    b. false

17. There is little difference between farm equipment and automobiles because the principles of operation are basically the same.

    a. true    b. false

## Multiple Choice

1. New laws demand that how many firefighters must be outside the structure as the first team of two goes in to fight the fire?

    a. one

    b. two

    c. three

    d. four

2. The text suggests that firefighters carry which of the following tools with them when approaching a structure fire? (choose two)

    a. flashlight

    b. forcible entry tool

    c. macloud

    d. water thief

3. If firefighters could not acquire a guideline, hose, or rope to assist in a search, which of the following could they use for search reference?

    a. instinct

    b. smoke output

    c. a wall

    d. furniture

4. One of the ways to identify the newer ropes used for guidelines is to look for which of the following features?

    a. smooth versus rough

    b. reflective materials

    c. webbing fiber

    d. rope weave

5. When attempting to escape a structure with no visibility, and following a hose, one is told to remember what about the hose?

    a. go wet to dry

    b. the harder, the closer to the pump

    c. vibrations

    d. male ends toward fire

6. Which of the following search methods is the most dangerous in fire conditions?

    a. crawl

    b. wall

    c. secondary

    d. primary

7. In a commercial fire during operating hours, the firefighter would most likely find victims hiding in which of the following areas? (choose two)

    a. closets

    b. offices

    c. restrooms

    d. exhibit floor

8. It is recommended that proper lifting technique be used in victim rescue. The body part most often used, injuring firefighters, is which of the following?

    a. legs

    b. arms

    c. back

    d. thighs

9. Firefighters use drags and carries in victim rescue. Which of the following is *not* one of the common carries utilized by firefighters?

    a. extremity carry

    b. seat carry

    c. firefighter's carry

    d. dead lift carry

10. The drag is another way of removing victims from danger. Which of the following drags is *not* one commonly associated with the fire service?

    a. blanket drag

    b. dead arm drag

    c. clothing drag

    d. sit and drag

11. If given the choice, which of the following modes of patient transport would be best for both the patient and the rescuer? (choose two)

    a. backboard

    b. blanket carry

    c. seat carry

    d. litter

12. All of the following choices are part of the steps involved in an extrication incident. Which one should firefighters consider first upon arrival on scene?

    a. patient removed

    b. scene assessment

    c. disentanglement

    d. vehicle stabilization

13. The larger air bags used for rescue deflate to as small as 1 inch and inflate up to how many inches?

    a. 6 inches

    b. 10 inches

    c. 20 inches

    d. 30 inches

14. In vehicle stabilization it is important to completely remove the weight of the vehicle from which of the following choices?

    a. wheels

    b. suspension system

    c. tires

    d. not important

15. Rescue harnesses are categorized into classes with which class being the one known as a "ladder belt"?

    a. Class 1A

    b. Class 2A

    c. Class 3A

    d. Class 4A

16. If a patient is going to be suspended in a harness for any reason it is highly recommended that the patient be placed in which of the following harnesses?

    a. Class 3

    b. Class 2

    c. Class 1

    d. web seat

17. Collapse scenes present special hazards. Which of the following are considered two of the most common hazards to firefighters? (choose two)

    a. primary collapse

    b. water piping damage

    c. secondary collapse

    d. gas leaks

18. Which of the following is *not* considered a type of structural collapse?

    a. pancake

    b. lean-to

    c. V-type

    d. flat

19. One of the most life-threatening situations during a trench collapse is which of the following?

    a. dirt around victim's chest

    b. wet dirt versus dry dirt

    c. dust cloud

    d. legs getting stuck

20. The most common hazard encountered at a confined space incident is which of the following?

    a. rich atmosphere

    b. toxic gases

    c. oxygen deficiency

    d. opening size

21. Which of the three harnesses shown in Figure 16-3 is a Class 1 harness?

    a. a

    b. b

    c. c

    d. none of the above

a.

b.

c.

**Figure 16-3** Identify the Class I harness.

22. All the following can help evaluate a rescue profile for an occupancy *except:*
    a. occupancy type/time of day
    b. fire and smoke conditions
    c. activity clues
    d. hydrant locations

23. In order to prevent musculoskeletal injuries when performing victim carries, the firefighter should:
    a. create a power-center using a "tight-core"
    b. relax back muscles
    c. utilize webbing to gain leverage
    d. utilize a two-in/two-out concept

24. Thermal imaging cameras allow firefighters to:
    a. replace standard search procedures and penetrate deeper in smoke-filled occupancies
    b. compare smoke densities for faster fire attack
    c. "see" heat energy through smoke-obscured environments
    d. monitor toxic gas levels within a tank

## Fill-in-the-Blanks

    a. cribbing                 b. V-type

    c. pancake                d. shoring

    e. lean-to

Structural collapse is not uncommon in today's fire service. One such collapse a few years ago involved a double-decker freeway in which the top layer collapsed for most of its length onto the lower portion. This (1) _____ collapse left only small voids in which cars were trapped. The (2) _____ collapse is very common in earthquakes when one side of the supporting wall or floor anchoring system fails. The last type of collapse involves the center of the floor or roof that becomes overloaded with rainwater or snow and drops. This is called a (3) _____ collapse and is also quite common. Two things firefighters use when entering these areas are (4) _____ , which involves systematically stacking lumber to hold up a structure, and (5) _____ , which uses timbers to support a load.

6. Describe the process of placing a patient on a backboard being used in Figure 16-4.

    a. _____

    b. _____

    c. _____

    d. _____

    e. _____

    f. _____

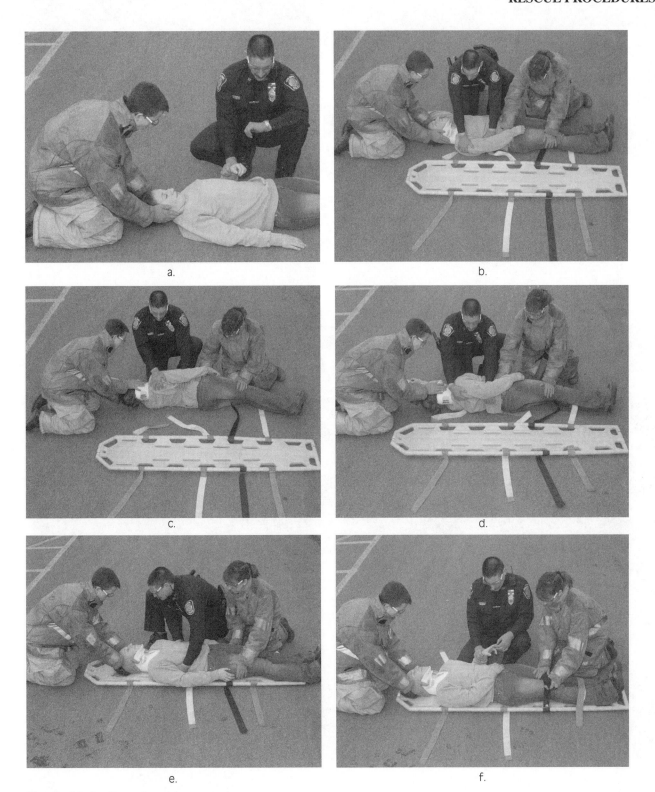

a.

b.

c.

d.

e.

f.

**Figure 16-4** Describe this operation.

# FORCIBLE ENTRY

# QUESTIONS

## Matching

Match the family or group of tools to the tools themselves.

      a. pike pole

      b. sledgehammer

      c. K tool

      d. crowbar

      e. bolt cutters

1. Striking tool: _____

2. Prying tool: _____

3. Cutting tool: _____

4. Pulling tool: _____

5. Through-the-lock tool: _____

6. Match the terms listed with the blanks in Figure 17-1.

| | |
|---|---|
| hinges | wall |
| jamb | knob-locking device |
| frame | door |

      a. _____

      b. _____

      c. _____

      d. _____

      e. _____

      f. _____

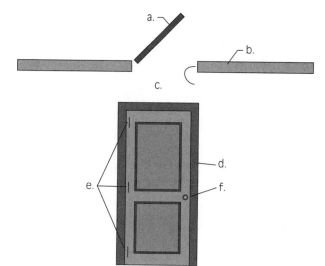

**Figure 17-1** The parts of a door assembly.

# True or False

1. Structural entry is important and must impede an aggressive attack to be effective.
    a. true        b. false

2. According to the text, as security concerns increase so will the number and types of locks.
    a. true        b. false

3. The primary concept of forcible entry is brute muscle rather than brainpower.
    a. true        b. false

4. It is logical to assume that a tool that is too heavy cannot be moved fast enough to develop proper striking force.
    a. true        b. false

5. Power hydraulic spreaders can be used for forcible entry in certain circumstances.
    a. true        b. false

6. While a firefighter may use a carbide-tipped blade for cutting composition roofing, it is advised that a regular blade be used because of binding problems with carbide.
    a. true        b. false

7. The use of cutting torches is common in the fire service; however, it requires specialized training.
    a. true        b. false

8. The wood handles of many tools must be kept in good condition. One way of doing this is to keep them sanded and painted or varnished.
    a. true        b. false

9. Building and life safety codes require revolving doors to collapse and allow occupants a rapid exit if needed.
    a. true        b. false

10. It is a good idea to coordinate all forcible entry operations with fire attack and ventilation.
    a. true        b. false

11. For doors that are hung in jambs with hinges, forcible entry is accomplished by working with the direction of swing.
    a. true        b. false

12. When carrying an ax, the text suggests that it is carried with the blade angled down and toward your outside foot so that the pick end is aimed away in case of a fall.
    a. true        b. false

# Multiple Choice

1. One of the first operations conducted before anything else at building fires is:
    a. secure utilities
    b. fire attack
    c. conduct forcible entry
    d. check every room

2. When performing forcible entry, it is important that which of the following is always attempted first?
   a. glass is broken out
   b. smoke ejectors are set up
   c. roof ventilation is accomplished
   d. try to just open the door

3. The flathead ax is available in different weights. Which of the following is the text's suggested optimum weight for striking purposes in forcible entry?
   a. 4 lb
   b. 6 lb
   c. 8 lb
   d. 10 lb

4. The flathead ax is most commonly used as which of the following?
   a. striking tool
   b. battering tool
   c. prying tool
   d. cutting tool

5. Name the two types of power cutting tool-saws. (choose two)
   a. rotary saw
   b. air chisel
   c. ax
   d. chain saw

6. The use of power saws is a two-firefighter operation. Name the two positions in this operation. (choose two)
   a. lead person
   b. saw operator
   c. guide firefighter
   d. backup operator

7. Which of the following is plate glass that has been heat-treated to increase its strength?
   a. tempered glass
   b. Plexiglas
   c. window glass
   d. regular glass

8. Which of the following is *not* one of the three methods of firefighter forcible entry?
   a. conventional
   b. specialized
   c. through-the-lock
   d. power tools

9. What type of glass breaks into very sharp, knife-like shards?
   a. tempered glass
   b. Plexiglas
   c. window glass
   d. regular glass

10. What are the two most common reasons for breaking out a window? (choose two)

    a. to increase view

    b. to release occupants

    c. to gain entry

    d. to create horizontal ventilation

11. The primary purpose of the tools in Figure 17-2 is for which of the following?

    a. prying

    b. pulling

    c. cutting

    d. striking

**Figure 17-2**  Name the primary purpose of these tools.

12. The primary purpose of the tools in Figure 17-3 is for which of the following?

    a. prying

    b. pulling

    c. cutting

    d. striking

**Figure 17-3** Name the purpose of these tools.

13. The best method of forcing entry through an opening with a tempered glass door is which of the following?
    a. break with a pick-head ax
    b. ram bar in the bottom corner
    c. rotary saw on the hinges
    d. through the lock

# Fill-in-the-Blanks

    a. building construction
    b. skills
    c. size-up
    d. knowledge
    e. locking devices

Forcible entry is a combination of (1) _____ and (2) _____ used to gain entry to a structure when the openings are locked. To perform this, firefighters must be knowledgeable in (3) _____ and the operation of (4) _____ . The firefighter must be able to (5) _____ the quickest and easiest way to gain entry to the building.

6. What type of saw utilizes the blades shown in Figure 17-4? Also, what are each of the blades shown in the figure designed to cut?

_____ saw is used for cutting _____ and _____ .

**Figure 17-4** Identify these blades.

7. What is the way *The Firefighter's Handbook* recommends for making entry into the type of door shown in Figure 17-5?

_____

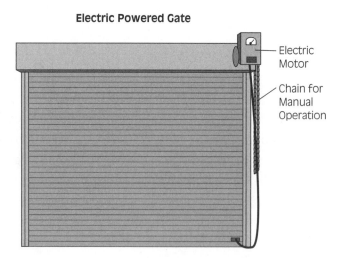

**Figure 17-5** Roll-down steel doors.

# CHAPTER
# 18

# VENTILATION

# QUESTIONS

## Matching

Match the correct term with the definitions provided.

      a. horizontal

      b. pressure

      c. halon

      d. carbon monoxide

      e. vertical

1. Ventilation can be defined as the planned release of _____ , heat, smoke, and gases.

2. _____ will take the place of oxygen in the blood when inhaled.

3. Vertical ventilation is the removal of gases through _____ avenues.

4. _____ ventilation permits the fire's by-products to be pushed out of the structure ahead of the fire attack team by using a window or doorway.

5. Many times ventilation is the wrong thing to do. One such time is when _____ has been used as an extinguishing agent prior to the fire department's arrival.

## True or False

1. Holding the fire's heat in the structure while lines are being set is a good practice that makes the attack easier.

    a. true      b. false

2. Entry into a heavy smoke condition can obscure light so completely that even a flashlight is rendered ineffective.

    a. true      b. false

3. The pressure a fire creates will many times spread that fire into other areas of a structure.

    a. true      b. false

4. A good indicator that the smoke and heat are meeting vertical movement resistance is a large volume of the smoke coming from around the window seams.

    a. true      b. false

5. There is really no difference between vertical and horizontal ventilation except in the equipment involved.

    a. true      b. false

6. An advancing hose stream will push smoke, heat, and steam out of the firefighter's way, causing no need for ventilation.

    a. true      b. false

7. Wind is another factor that can dramatically alter ventilation.

    a. true      b. false

8. During flashover conditions the best skill a firefighter can have is that of recognition and avoidance.

    a. true      b. false

9. Positive pressure ventilation is a great post-fire procedure; however, it has not been perfected enough to use during the fire.

    a. true      b. false

10. When using the ax for venting glass, the side of the ax is used to actually break the glass.

    a. true      b. false

11. Tests have shown that effective pressures can be generated up to twenty-five stories from the location of a properly placed PPV blower.

    a. true      b. false

12. The proper practice of cutting a hole in a roof is to quickly and accurately cut deeply all the way around the hole, leaving nothing in the way.

    a. true      b. false

13. The good firefighter will always keep the ladder nearby when operating on a roof, since that ladder will be the only way of escape.

    a. true      b. false

14. Research has shown that one large hole is much more effective than several smaller holes.

    a. true      b. false

15. In order to ensure good horizontal ventilation on all levels of an occupancy, it is good practice to open every window about one-third from both the top and bottom.

    a. true      b. false

16. HVAC, better known as horizontal and vertical air control, is most commonly used for ventilation purposes in high-rise firefighting.

    a. true      b. false

17. Misuse of positive pressure ventilation can severely accelerate fire growth.

    a. true      b. false

## Multiple Choice

1. The heat emitted by the fire will carry which of the following items throughout the structure?

    a. steam

    b. ventilation

    c. attack

    d. smoke

2. When fire burns, air heats, expands, and: (choose two)

    a. cools

    b. slows

    c. rises

    d. becomes lighter

3. When water is applied to a fire and the fire is extinguished, what takes place with the heat? It:

    a. expands

    b. cools

    c. rises

    d. smokes

4. When water is applied to fire and it turns to steam, there is an expansion of how much?
   a. 1,200 times
   b. 1,700 times
   c. 2,100 times
   d. double

5. If horizontal ventilation is performed at the wrong time or the wrong place, what can the result be?
   a. accelerated fire
   b. windows will break
   c. vertical ventilation
   d. smoke change

6. The removal of a window facing the wind during ventilation can result in which of the following?
   a. increased ventilation
   b. interior air movement reversal
   c. successful product extinguishment
   d. the fire going out

7. A humid, rainy, or foggy day will cause the smoke-laden air to do what as it is ventilated?
   a. get wet
   b. lift
   c. drop
   d. evaporate

8. As the trapped heat from the fire collects at the ceiling, what happens to it? It: (choose two)
   a. gets hotter
   b. gets cooler
   c. mushrooms
   d. gets heavier

9. According to the text, the survival time of a fully dressed firefighter (PPE) in a flashover would be:
   a. ten to fifteen seconds
   b. twenty to thirty seconds
   c. one minute
   d. five minutes

10. Of the following choices, which one is *not* a sign of a potential backdraft?
    a. smoke-stained windows
    b. smoke pushing or puffing out
    c. cooled interior conditions
    d. a tightly sealed building or room

11. Which of the following choices would best describe simply opening doors and windows during a fire?
    a. vertical ventilation
    b. positive pressure ventilation
    c. reverse ventilation
    d. natural ventilation

⊞ 12. In climate-controlled buildings, such as high-rise occupancies, what is the system the fire department can use to mechanically remove smoke and combustion products?

    a. skylights

    b. HVAC

    c. fire pumps

    d. vestibule doors

13. Using a fan is a main ingredient in assisting with positive pressure ventilation. Which way does the fan face and where is it usually placed?

    a. away from building and in window

    b. away from building and near doorway

    c. toward building and near doorway

    d. not used in PPV

14. What is the term used for ventilation when a hose stream is used?

    a. vertical

    b. positive

    c. negative

    d. hydraulic

15. Of the following tools, which one, when used for ventilation, can pose the most danger to the firefighter?

    a. fan

    b. pike pole

    c. ax

    d. hose stream

16. Fighting fire in basements or cellars is difficult and hard work. Using a fan to pull smoke from the area by placing the fan over the opening is used primarily for which of the following?

    a. overhaul

    b. attack

    c. PPV

    d. HVAC

17. When cutting holes in the roof for vertical ventilation, where is the best place for the primary hole?

    a. at the corners

    b. at the ridge

    c. at the ventilator

    d. over the fire

18. What is the term used for cutting the roofing material and not the rafters, then pulling the roofing material up by rotating it on the rafters?

    a. slip cut

    b. trench cut

    c. louver cut

    d. big hole cut

19. Of the following types of roof cuts, which two are most usually considered defensive? (choose two)
    a. primary
    b. strip cut
    c. secondary
    d. trench cut

20. Venting too early can cause which of the following to occur?
    a. fire extension
    b. backdraft
    c. flashover
    d. quick extinguishment

21. Figure 18-1 depicts two types of ventilation commonly used by firefighters. Which of the following are these two types of ventilation? (choose two)
    a. venturi
    b. evacuation
    c. positive pressure
    d. hydraulic

22. When discussing types of ventilation, the text describes cutting a hole in a roof as what type?
    a. mechanical
    b. natural
    c. horizontal
    d. physical

## Fill-in-the-Blanks

    a. rollover
    b. vertically
    c. backdraft
    d. flashover
    e. horizontally

When the fire has been in the free-burning phase for some time and everything in the confined area ignites at almost the same time, a (1) _____ can occur. The best way to avoid this is to quickly cool the upper area of the space or (2) _____ ventilate. The best way to tell if this phenomenon is about to occur is to watch for (3) _____ .

The second phenomenon that often takes place at an interior fire during the smoldering or last stage of fire is called a (4) _____ . The best way to avoid this occurrence is to know the signs and symptoms and to quickly (5) _____ ventilate.

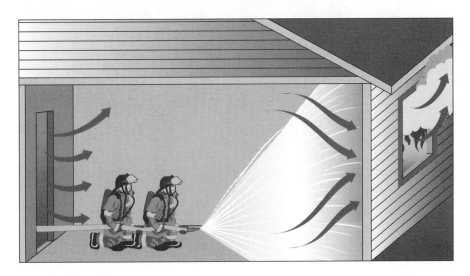

**Figure 18-1** Name the ventilation types.

6. Figure 18-2 shows a building with three types of holes cut into the roof. Name each of the types of holes.

a. _____

b. _____

c. _____

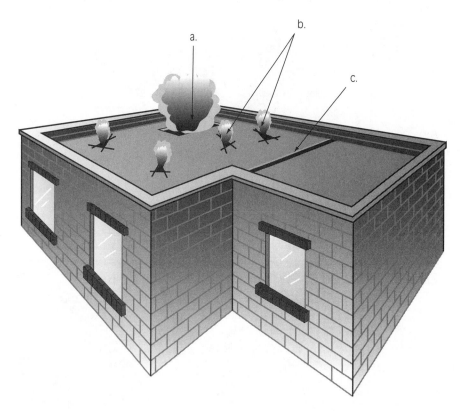

**Figure 18-2**   A building with holes cut in the roof.

7. Name the process shown in Figure 18-3 and describe how it works.

_____

_____

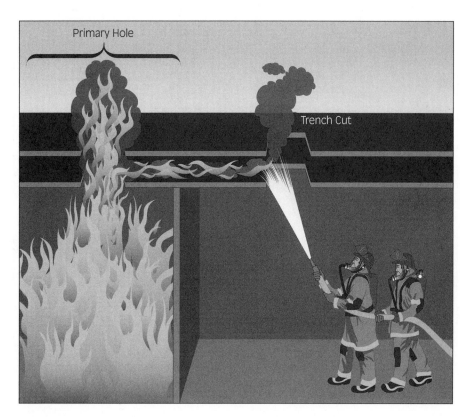

**Figure 18-3** Describe this operation.

8. Briefly describe the operation shown in Figure 18-4.

_____

_____

**Figure 18-4** Identify this operation.

# FIRE SUPPRESSION

*Courtesy of Central Net Fire*

# QUESTIONS

## Matching

Match the correct term with the definitions provided.

     a. control room

     b. exposure

     c. rate of spread

     d. flammable range

     e. fire intensity

     f. BLEVE

1. The speed that fire runs through grassy or light brush areas is termed _____ .

2. The _____ is commonly associated with the amount of heat and flame given off in a wildland fire, usually in heavier fuels.

3. When a gas is too rich or too lean to burn, it is associated with the fuel's _____ .

4. In a high-rise fire there is a location where firefighters can go in order to see what is going on with the building systems. This room is called the _____ .

5. When a fire is threatening something, it is commonly called a(n) _____ , and thus it will be given a priority over the actual burning materials themselves.

6. Using the preceding terms, what danger is posed by the fire in Figure 19-1? _____

**Figure 19-1** Water is applied to the heated metal surface.

7. Proper stream placement is important when trying to defend an exposed structure from radiated heat. Which of the two drawings in Figure 19-2 is the correct positioning of this water stream? Circle one:  a  or  b

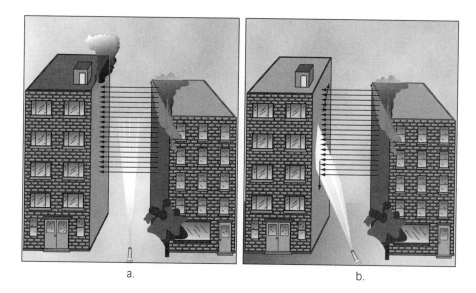

a.                                          b.

**Figure 19-2**  Two drawings of fire exposures. Which water application will stop an exposure fire?

8. Match the best extinguishing method with the fires described below:

a. foam                1.        Small flammable liquid fires
b. let it burn          2.        Larger fires of pooled flammable liquids
c. $CO_2$              3.        Propane tanker venting due to fire exposure

# True or False

1. It is well known that the earlier the fire department arrives on scene and the earlier that the fire suppression begins, the higher the fire losses in saving the affected structure.
   a. true        b. false

2. The application of water as an extinguishing agent is perhaps the simplest form of fire suppression, since it works on all fires.
   a. true        b. false

3. Resources, while important, do not have much effect on fire suppression because fire attack will not differ. The concept of fire extinguishing never changes.
   a. true        b. false

4. The larger campaign fires are usually wildland or ground cover fires.
   a. true        b. false

5. Of the three components of the wildland fire triangle, weather is possibly the most dynamic of all.
   a. true        b. false

6. Heavier fuel in wildland environments is usually associated with high rates of spread as opposed to lighter grasses and brush.

    a. true     b. false

7. When heated, the hydraulic fluid-filled bumper systems on many cars will blow a fuse, allowing the pressure to escape safely.

    a. true     b. false

8. When rating the toxicity of different materials in fire, the order from least to most toxic is: plastic, wood, and cloth.

    a. true     b. false

9. The best way to disarm an air bag system in an automobile would be to disconnect the car's battery.

    a. true     b. false

10. Flammable liquid fires can actually be complicated by improper actions of firefighters.

    a. true     b. false

11. A liquid that is water soluble is one that will normally float on water.

    a. true     b. false

12. One of the first things a fire officer must do when arriving at an incident is to create a plan of attack.

    a. true     b. false

13. Most fire departments have SOPs that determine who is first on scene and where they should be located.

    a. true     b. false

14. According to the text, the best size line to pull on a fire is the largest size needed rather than trying to "go large" later.

    a. true     b. false

15. When considering tactical objectives, it is important to accomplish them in order, since the safety of firefighters, occupants, and others is of primary importance.

    a. true     b. false

16. Before anything else, the fire must be extinguished in order to accomplish other objectives such as rescue and overhaul.

    a. true     b. false

17. The term "under control" means that the fire has been extinguished and everything is now under control.

    a. true     b. false

18. The two in/two out rule is not in force if a known rescue is needed.

    a. true     b. false

19. The chances of having to save lives are less in business and mercantile occupancies compared with residential occupancies.

    a. true     b. false

⊞ 20. Resources needed at a multistory fire will be greater than at a single story with similar occupancy.

    a. true      b. false

21. During the overhaul of large, high-piled storage fires outdoors, the use of SCBA is not needed.

    a. true      b. false

## Multiple Choice

1. Most structural firefighting will involve the suppression of which of the following classes of materials?

    a. Class A

    b. Class B

    c. Class C

    d. Class D

2. What do many firefighters use to determine the stability of a structure that is on fire?

    a. fire load

    b. flame length

    c. fire start time

    d. suppression technique

3. The wildland fire triangle differs from the structural triangle primarily in which area?

    a. heat

    b. weather

    c. fuel

    d. oxygen

4. Of the wildland fire triangle features, which one is the reason for most wildland firefighting deaths?

    a. heat

    b. weather

    c. fuel

    d. oxygen

5. In the case of fuels in the wildland environment, it can be said that which of the following is most true? (choose two)

    a. lighter fuels burn faster than heavy

    b. lighter fuels burn hotter than heavy

    c. heavy fuels burn faster than lighter

    d. heavy fuels burn hotter than lighter

6. Of the following topographic features, which one is *not* considered a "watch out" feature?

    a. saddle

    b. ridge

    c. drainage

    d. box canyon

7. What exterior portion of a burning vehicle is the least safe to stand near?
   a. doors
   b. trunk
   c. gas tank
   d. bumpers

8. The fully involved automobile engine will present the firefighter with all but which of the following classes of fire?
   a. Class A
   b. Class B
   c. Class C
   d. Class D

9. The interior of an automobile is very toxic when on fire. What is perhaps the second most dangerous item in an automobile's interior?
   a. dash
   b. windows
   c. hatchback struts
   d. electronics

10. According to the text, which of the following can be considered a flammable liquid's most hazardous profile?
   a. vapors
   b. temperature
   c. viscosity
   d. flow rate

11. Flammable gases are typically stored in what fashion? (choose two)
   a. liquid
   b. heated
   c. frozen
   d. pressurized

12. Which of the following items concerning a flammable product will determine what the responder should wear for survival?
   a. time of day
   b. length of burn
   c. toxicity
   d. location of hazard

13. It must be understood that fire has how many sides that must be considered?
   a. two
   b. four
   c. six
   d. eight

14. Which of the following is not an attack method?
    a. indirect
    b. direct
    c. select
    d. combination

15. In the acronym RECEO, the "C" stands for:
    a. control
    b. confinement
    c. conservation
    d. consideration

16. Another way of setting fireground priority is by using all of the following benchmarks except:
    a. fire control
    b. property conservation
    c. rescue
    d. ventilation

17. When considering modes of fire attack, such as offensive or defensive, one must understand that these are based primarily on which of the following?
    a. flame length
    b. resource adequacy
    c. rescue
    d. salvage

18. If a team were to pull a smaller line than is needed to extinguish the fire and then use it to hold the fire during rescue, what type of attack is this?
    a. offensive
    b. direct
    c. tactical
    d. defensive

19. The OSHA rule on two in/two out applies to all but which of the following situations?
    a. confined space
    b. firefighting
    c. HAZMAT response
    d. confirmed interior rescue

20. What side of the fire does the knowledgeable firefighter work from in interior firefighting?
    a. top side
    b. front side
    c. unburned side
    d. hottest side

▥ 21. In which phase of the operation does the cause of the fire become most apparent?

    a. attack

    b. overhaul

    c. salvage

    d. rescue

22. Which of the following types of fires results in the greatest loss of firefighters' lives?

    a. mercantile

    b. education

    c. assembly

    d. residential

23. According to the text, what is the most common element in residential firefighting?

    a. layout

    b. hazards

    c. occupant locations

    d. size of fires

▥ 24. In most structure fires, unless something is prearranged, where does the first-due engine locate itself?

    a. the rear

    b. the side

    c. the front

    d. the smokiest area

25. In a high-rise fire, firefighters can use the elevator more safely if which of the following are true? (choose two)

    a. elevator has fireproof doors

    b. elevator has firefighter lock-out controls

    c. elevator has computer programming

    d. elevator is part of a split bank

26. When laying hose in the stairwell, it is best to lay it from which of the following locations?

    a. straight into the fire floor for direct attack

    b. from the floor above for better hose pullability

    c. from the floor below for safety

    d. from the engine below

▥ 27. According to the text, what is the key to fighting a basement or cellar fire safely?

    a. ventilate quickly

    b. foam generation

    c. seal the room

    d. flood the room

28. In a sprinklered building with fire in the building, what effect will the sprinklers have on the fire besides partially extinguishing it?
    a. make the floor slippery
    b. wet down the firefighters
    c. complicate radio traffic
    d. hold down the smoke

FFI 29. The best method of protecting an exposure from radiated heat is to do which of the following?
    a. apply a water curtain
    b. attack the main fire
    c. close all windows
    d. apply water to the exposure

FFI 30. Controlling a wildland fire is best described by which of the following?
    a. extinguishing it
    b. surrounding it
    c. doing water drops
    d. backfiring it

31. Where should firefighters place the fire engine when approaching an accident in the #3 lane of the freeway?
    a. beyond the accident in the #3 lane
    b. before the accident in the #3 lane
    c. halfway in the #2 and #3 lanes before the accident
    d. halfway in the #3 and #4 lanes before the accident

## Fill-in-the-Blanks

    a. largest group
    b. opposing streams
    c. anyone else
    d. high-rise
    e. contents
    f. exposed areas
    g. indirect
    h. closest to the fire
    i. extinguishment
    j. direct

Most structure fires begin in (1) _____ and spread to the structure itself. Upon arrival the first-in attack team may consider either a(n) (2) _____ attack on the seat of the fire, or they may consider doing a(n) (3)_____ attack if they cannot see the seat of the fire. Many times when fighting structure fires not enough water is used in an attempt to create less damage, and the firefighter may find that (4) _____ did not take place. This can prove disastrous. On the other hand, if too much water is applied it can come from (5) _____ , causing an unsafe attack by the firefighting teams.

If a rescue is needed in an occupancy during a fire, the fire service has set priorities for who is rescued first. In order of importance when time is of the essence, those (6) _____ are brought out first, then the (7) _____ are brought out next in order. The group that would follow this would be (8) _____ , and then lastly those in (9) _____ would be saved. It is a fact that in (10) _____ occupancies the number of firefighters needed to accomplish these rescues would be significantly greater.

11. Fill in as many blanks as you can in Table 19-1 regarding the fireground factors that firefighters would consider when faced with a structure fire as they go on scene.

## Fireground Factors

**BUILDING** _____

_____

_____

_____

**FIRE** _____

_____

_____

_____

**OCCUPANCY** _____

_____

_____

**Table 19-1**   There can be numerous building fires and occupancy factors to consider.

12. The wildland fire triangle differs from the structural fire triangle. Fill in the blanks in the wildland triangle shown in Figure 19-3.

**Figure 19-3** The wildland fire triangle differs from the structural fire triangle.

13. What are some common characteristics of the fire shown in Figure 19-4?

**Figure 19-4** Identify the common characteristics found in this type of fire.

14. Describe the operation being shown in Figure 19-5.

**Figure 19-5** Describe this operation.

15. Figure 19-6 shows a fire in the rear bedroom of this dwelling. Draw your first attack line on this figure.

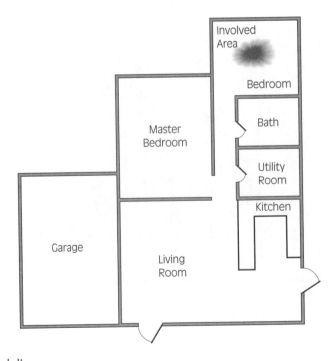

**Figure 19-6** Draw the attack line.

16. If one were to come upon the fire in Figure 19-7, how would the rescue of these occupants be prioritized? Please write numbers 1 through 4 under each set of victims.

**Figure 19-7** When numerous rescues face the first team, they must prioritize their actions based on victim exposure.

17. Figure 19-8 shows two wildland fire operations. One is a direct attack and the other an indirect attack. Place the correct type under each photo.

a._____

b._____

**Figure 19-8** Identify the attacks shown.

# SALVAGE, OVERHAUL, AND FIRE CAUSE DETERMINATION

# QUESTIONS

## True or False

1. It is said that the mainstay of salvage is the salvage cover or variation of the salvage cover.
   a. true    b. false

2. The outer edge of the salvage cover is ringed with grommets spaced no farther than 12 inches apart for ease of tie-down and hanging.
   a. true    b. false

3. The floor runner is really a form of salvage cover, since it is used mostly to protect the flooring from hose abrasions.
   a. true    b. false

4. Salvage is basically working with large covers and very little use of tools because the covers are designed to cover everything.
   a. true    b. false

5. In some cases when the cover is exposed to hazardous materials, it is good practice to wash it with a mild acid solution and then soap to remove the chemicals.
   a. true    b. false

6. The covers are highly water-resistant so after washing they can be inspected and then folded for use again.
   a. true    b. false

7. After using a chain saw on an incident it is important to take it apart and clean it thoroughly.
   a. true    b. false

8. The salvage crew follows the attack crew after the fire has been knocked down. They must wear PPE but do not need SCBA because the smoke has been removed by the other team.
   a. true    b. false

9. One of the most common hazards of salvage work is ceiling collapse.
   a. true    b. false

10. The only way to remove water from an upstairs floor is to vacuum it out with a water vacuum.
    a. true    b. false

11. The best way to salvage larger items is to cover them in place, since removing them may be too difficult.
    a. true    b. false

12. It is a common concept that many times water damage can be worse than the fire damage.
    a. true    b. false

13. When creating a water chute, the salvage cover is placed outside of a ladder so that water is channeled to another location.
    a. true    b. false

14. Many times overhaul can cause more damage to an occupancy than the fire by breaching walls and ceilings.

    a. true      b. false

15. The best thing about residences with blown-in insulation is that the insulation is fireproof and the firefighter will not have to wet that area or care about it as much as other areas.

    a. true      b. false

16. The primary role of a company officer includes small safety-oriented tasks when necessary.

    a. true      b. false

17. Salvage operation plans should consider what the homeowner would want to save.

    a. true      b. false

# Multiple Choice

1. Many times items requiring salvage are not common furniture. Two such items would be which of the following? (choose two)

    a. couch

    b. bed

    c. patio furniture

    d. automobile

2. Which of the following choices are the basic premise of salvage operations? (choose two)

    a. removing all burned material

    b. locating the fire's origin

    c. removing harmful atmospheres

    d. covering furnishings and valuables

3. Which of the following materials is *not* commonly used to make salvage covers?

    a. canvas

    b. cloth

    c. plastic

    d. treated canvas

4. Which of the following answers is a cover that has advantages over others in water resistance and weight?

    a. canvas

    b. cloth

    c. plastic

    d. treated canvas

5. When a salvage cover is folded into itself so that material is caught and trapped inside, it is said to be folded in which fashion?

    a. loose fold and roll

    b. butterfly fold

    c. capture roll

    d. cross-fold roll

6. According to the text, when placing a number of folded salvage covers in a stack it is a good idea to do what in order to separate them easily?

    a. place them crossways every other time

    b. put a paper towel between each

    c. paint the edges of every other cover

    d. there is no known way

7. Which of the following answers is *not* a recommended way of quickly salvaging valuable materials?

    a. covering the item

    b. moving the item away

    c. rinsing the item quickly

    d. removing harmful substances from the area

8. The most common warning sign of water leaking through a ceiling with potential failure is:

    a. water from an electric outlet

    b. sweating on the wall

    c. sagging ceiling and wall studs

    d. seams showing as water seeps into them

9. The most common tool used by the fire service for stopping the flow of water from a sprinkler head is a:

    a. wedge

    b. plug

    c. vice

    d. spanner

10. When moving items about a room for salvage, what does the firefighter look for first?

    a. the largest item

    b. the smallest item

    c. the most valuable item

    d. the wettest item

11. Which of the following is *not* a common salvage cover throw?

    a. counter payoff

    b. roll out

    c. shoulder toss

    d. balloon toss

12. The fire service commonly uses two methods of water damage prevention when large amounts of water are a problem. It is either caught or sent elsewhere with salvage covers using which of the following two common methods? (choose two)

    a. catch-all

    b. water slide

    c. relief basin

    d. water chute

13. Which of the following is not really considered a common salvage tool?

    a. Halligan tool

    b. pike pole

    c. shovel

    d. rubbish hook

14. It is not uncommon to have fire travel through walls using this method. (choose two)

    a. convection

    b. conduit

    c. radiation

    d. ductwork

15. A good rule is to revisit the burned occupancy two times after the fire has been extinguished. The text suggests which intervals for this activity?

    a. at two and ten hours

    b. at four and eight hours

    c. at six and ten hours

    d. at eight and twelve hours

16. Identify the operation shown in Figure 20-1.

    a. salvage fold

    b. over throw

    c. shoulder toss

    d. balloon toss

**Figure 20-1** Identify this operation.

17. Identify the operation shown in Figure 20-2.
  a. salvage fold
  b. cover deployment
  c. shoulder toss
  d. balloon toss

**Figure 20-2** Identify this operation.

18. The basic premise of salvage operation is:
  a. to remove the harmful atmosphere from the material or to protect the material from the atmosphere.
  b. to protect exposures
  c. to push toxic gases into a central location with positive pressure
  d. to ensure all exits are operational for rapid evacuation of salvage crews

# CHAPTER
# 21

# PREVENTION, PUBLIC EDUCATION, AND PRE-INCIDENT PLANNING

# QUESTIONS

## True or False

1. The fire prevention officer plays one of the key roles in ensuring that the department meets its goals of preserving life and property in the community.
   a. true     b. false

2. A department that has the luxury of a fire prevention staff usually does not have any fire prevention responsibilities for the other officers.
   a. true     b. false

3. The purpose of a fire company inspection program is for both future prevention purposes and immediate results, since the main goal is to prevent fires.
   a. true     b. false

4. When inspecting in the field, the best method is to break up the locations in order to "surprise" the occupants.
   a. true     b. false

5. Out of courtesy and respect to building occupants, it is better not to inspect occupancies during a busy workday.
   a. true     b. false

6. The best approach to an inspection is to enter from the rear and visit the office last, since the owner will not have time to prepare the occupancy prior to the inspection.
   a. true     b. false

7. The unescorted crew will have more leeway in looking into "hidden" spaces, so it is best not to ask for help until violations are found.
   a. true     b. false

8. A good general rule, although it does not apply to all occupancies, is to require that exits be identified with an exit sign.
   a. true     b. false

9. Panic hardware is required in all assembly, educational, and institutional occupancies.
   a. true     b. false

10. Fixed extinguishing systems are required whenever cooking processes take place, with the exception of pizza and bread ovens.
    a. true     b. false

11. If an occupancy being inspected has flammable liquids on hand in small quantities under 5 gallons, the liquids are not regulated because the codes apply to only larger quantities.
    a. true     b. false

12. When inspecting a small business, it is possible to find extension cords being used for small back room appliances. This is allowed if the appliances are unplugged at night.
    a. true     b. false

▥ 13. As long as oily rags are stored or discarded in approved safety cans they can be kept outside the occupancy.

    a. true    b. false

▥ 14. An inspecting firefighter who finds a violation that is an immediate life hazard may demand that it be corrected before leaving the occupancy.

    a. true    b. false

15. National statistics show that almost all burn injuries and deaths come from residential occupancies; however, the majority of inspection programs focus on the business community.

    a. true    b. false

16. The owner of some types of businesses may be allowed to install doors that will not unlock for up to fifteen seconds after the door lock mechanism is activated.

    a. true    b. false

17. When making an appointment for an inspection, it is important to tell the business owner that the appointment may be cancelled without notice and a contingency appointment should be discussed.

    a. true    b. false

## Multiple Choice

1. Which of the following is *not* one of the three Es of fire prevention?

    a. education

    b. engineering

    c. enforcement

    d. execution

2. Of the Es listed, which one would deal with the interaction with architects, builders, and construction plans review?

    a. education

    b. engineering

    c. enforcement

    d. execution

▥ 3. Which two of the following tools are most likely found on the firefighter's/inspector's checklist for fire prevention inspection purposes? (choose two)

    a. shovel

    b. fire code reference manual

    c. clipboard with pad and pencil

    d. pry bar

▥ 4. Before leaving the station on inspections, it is suggested that the firefighter consider reviewing the inspection files and look for all but which of the following?

    a. past violations

    b. the legal owner's name

    c. past corrections made

    d. past compliance records

5. What is the best reason for including all members of the company on the inspection?

    a. more eyes to look for violations

    b. more code knowledge

    c. not everyone can be fooled

    d. building familiarization

6. If all attempts to obtain code compliance fail, the inspector must obtain which of the following in order to achieve compliance?

    a. permission from the chief to shut them down

    b. an administrative warrant

    c. insurer's guarantee of coverage

    d. building owner's release of liability

7. The order of importance for inspecting an occupancy is:

    a. front to rear

    b. top to bottom

    c. rear to front

    d. does not matter

8. According to the text, what is the most often neglected system in a building?

    a. exits

    b. flammable liquid storage

    c. storage height

    d. extinguisher availability

9. Of the following classifications of doors, which is the only one permitted to be used for a required exit?

    a. revolving

    b. overhead

    c. sliding

    d. swinging

10. Assembly occupancies require panic hardware on the doors, and the mechanism must operate with no more than how much force?

    a. 5 lb

    b. 10 lb

    c. 15 lb

    d. 50 lb

11. Inspecting fire extinguishers requires checking all but which of the following items?

    a. location in building

    b. proper rating

    c. pressure gauge showing charged

    d. date of container manufacture

FFI 12. Inspecting sprinkler systems requires checking all but which of the following items?

    a. water clarity

    b. FDC free of obstructions

    c. access to risers open

    d. pressure gauges show proper pressures

FFI 13. Inspecting fixed gaseous extinguishing systems requires checking all but which of the following items?

    a. access to manual discharge valves is open

    b. discharge nozzles are unobstructed

    c. cylinder gauges show proper pressure

    d. gas is less than five years old

FFI 14. Outside liquid storage containers or tanks must have a curb around the bottom to contain spilled liquid. What is that area called?

    a. dike area

    b. spill pool

    c. secondary containment

    d. flammable storage area

FFI 15. Which two of the following items may require immediate correction by the inspector? (choose two)

    a. extinguisher missing

    b. exit sign lights out

    c. welding near open flammable liquids

    d. blocked exits

16. According to the text, firefighters should remind homeowners that their chances of survival are how much greater if they have a smoke detector?

    a. 25 percent

    b. 50 percent

    c. 75 percent

    d. 100 percent

17. A good public presentation should contain all but which of the following?

    a. preparation

    b. presentation

    c. personality

    d. practice

18. Public education programs usually take different forms. Which of the following are two of the most common forms? (choose two)

    a. public service announcements

    b. school programs

    c. fire training

    d. investigative reporting

19. What kind of public training would be required when talking about bomb threats and natural disasters that would also apply to fire dangers?

    a. stop, drop, and roll

    b. prevention tactics

    c. emergency evacuation

    d. combustible hazards

20. One type of inspection that requires a greater response and greater preparation during an emergency is for which of the following?

    a. grocery stores

    b. chemical hazards

    c. dining establishments

    d. target hazards

21. What is wrong with the exit corridor in Figure 21-1?

    a. ceiling too high

    b. storage in corridor

    c. exit width too narrow

    d. wrong sprinkler system

**Figure 21-1** Exit corridors.

22. Which type of door represented in Figure 21-2 can be used as a required exit?

    a. overhead

    b. swinging

    c. sliding

    d. revolving

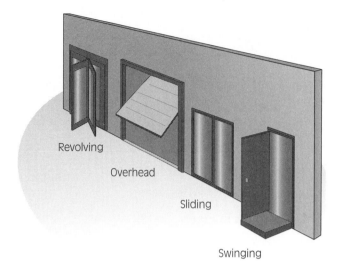

**Figure 21-2**   Four types of doors.

23. What is the common recommendation for how often to change the smoke detector batteries as shown in Figure 21-3?

    a. monthly

    b. every four months

    c. when clocks are reset for Daylight Savings

    d. as needed

**Figure 21-3**   Smoke detector battery installation.

24. The NFPA 704 placard system:
    a. uses colors to give first responders vital information
    b. uses numbers to give the first responders vital information
    c. uses colors and numbers to give first responders vital information
    d. uses drawings and numbers to give first responders vital information

# EMERGENCY MEDICAL SERVICES

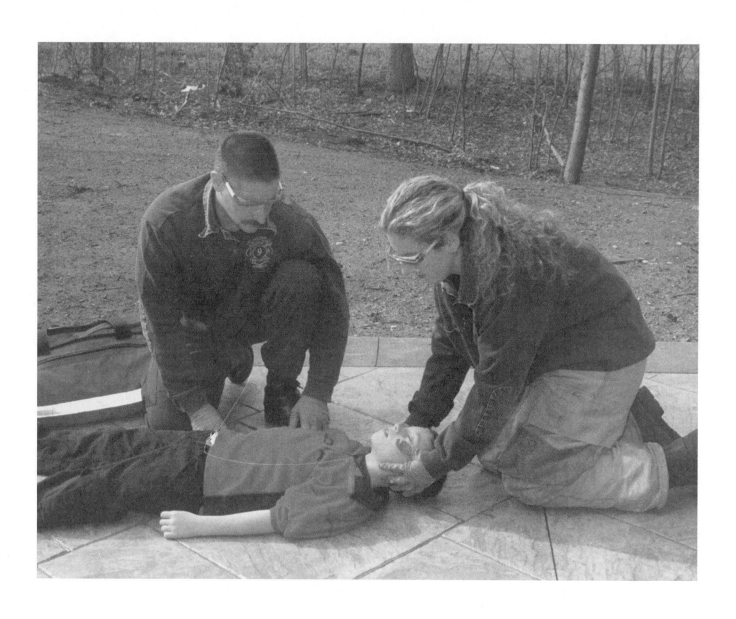

# QUESTIONS
## True or False

1. Overall, medical responses constitute more than 50 percent of total emergency responses for many fire departments across the country.
   a. true      b. false

2. Safety for firefighters in medical emergencies begins from the first step onto the engine until the end of the call after all equipment is cleaned and replaced on the unit.
   a. true      b. false

3. According to the law, medical responders must withhold medical treatment from minors until parental permission is given.
   a. true      b. false

4. Patients who are mentally capable and conscious have the right to refuse treatment, no matter how badly they are injured.
   a. true      b. false

5. If given a patient who is difficult to manage, the emergency responder may refuse to continue treatment, leaving the scene and patient to family or others.
   a. true      b. false

6. Upon entering an emergency scene and finding a dangerous situation, the responders may leave the scene until it is stabilized and safe to reenter.
   a. true      b. false

7. Body substance isolation (BSI) applies to every patient, with no exceptions.
   a. true      b. false

8. Gloves are to be worn for all patients displaying blood or body fluids but are not necessary for those who do not.
   a. true      b. false

9. Firefighters who regularly participate in emergency medical incidents should be immunized for hepatitis B.
   a. true      b. false

10. An automated external defibrillator is a portable computer-driven device that analyzes a patient's heart rhythm and delivers defibrillation shocks when necessary.
    a. true      b. false

## Multiple Choice

1. The medical responder of today may arrive on scene by a number of different modes. Which of the following is a typical mode of transportation for the medical responder?
   a. fire engine
   b. police car
   c. personal vehicle
   d. all of the above

2. The primary responsibility for firefighters responding to a medical emergency is:
   a. quick patient access
   b. vehicle stabilization
   c. to ensure responder safety
   d. to acquire needed resources

3. _____ is a legal term that means for every emergency medical incident, an emergency responder should treat the patient in the same manner as another emergency responder with the same training.
   a. consent
   b. triage
   c. liability
   d. standard of care

4. What type of burn affects not only the skin structure but also the tissues and muscles underneath?
   a. full thickness burn
   b. superficial burn
   c. partial thickness burn
   d. none of the above

5. At what point can the medical responder begin to relax standards of safety?
   a. receipt of alarm
   b. en route to call
   c. during emergency
   d. after tools are cleaned and put away

6. Which of the following is of great concern to the first responder when dealing with communicable diseases?
   a. HIV
   b. hepatitis B
   c. tuberculosis
   d. all of the above

7. When considering bloodborne pathogens and precautions, which of the following items should be a part of the firefighter's preventative barriers on every call? (choose two)
   a. gloves
   b. goggles or safety glasses
   c. helmet
   d. boots

8. The protective gloves worn by firefighters on medical emergencies should be changed how often?
   a. daily
   b. weekly
   c. before cleaning
   d. after each use

9. What is the term used to identify the first patient assessment, which is checking basic signs of life and life-threatening injuries?
   a. basic check
   b. initial assessment
   c. vital assessment
   d. medical check

10. As the patient is being assessed, it is common practice to recheck the vital signs often. About how often does the text recommend?
    a. every two to three minutes
    b. every three to five minutes
    c. every five minutes
    d. whenever symptoms change

11. Which of the following is *not* one of the common types of external bleeding?
    a. perfusion bleeding
    b. arterial bleeding
    c. venous bleeding
    d. capillary bleeding

12. The treatment for minor internal bleeding is:
    a. a tourniquet
    b. a pressure bandage
    c. pressure point application
    d. a cold compress or ice pack

13. Which of the following is *not* recommended as a method for dealing with a patient experiencing external bleeding?
    a. elevation of site
    b. suction
    c. pressure on pressure points
    d. exerting site pressure

14. A condition resulting in a decreased level of oxygen and nutrients to the body's tissues is:
    a. arterial bleeding
    b. hyperventilation
    c. perfusion
    d. shock (hypoperfusion)

15. Identify two common treatments for shock. (choose two)
    a. ensure airway and breathing
    b. raise legs
    c. place cold compresses on head
    d. exert external pressure

16. The firefighters in Figure 22-1 are attempting to accomplish which of the following EMS actions?

    a. stretch the neck

    b. establish an airway

    c. correct a dislocation

    d. stabilize the head and neck

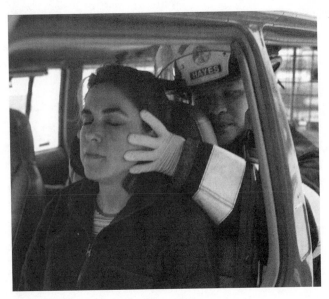 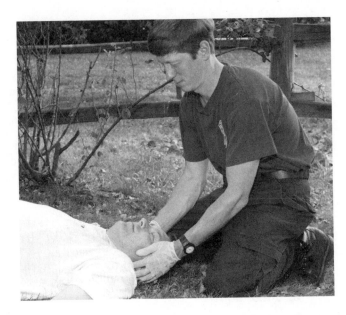

**Figure 22-1** Identify this activity.

## Fill-in-the-Blanks

    a. abrasion

    b. puncture

    c. laceration

    d. amputation

    e. avulsion

There are several common wounds the trained firefighter may treat in the field with bandaging. The (1) _____ wound is a common one involving an object that has entered the body at a specific point and then was retracted. An example would be a child being stabbed with a pencil by another student. Another one is the injury in which the skin is actually torn away leaving a flap or loose hanging piece. This is called a(n) (2) _____ and is bandaged with a sterile wrap. The (3) _____ is a wound in which the skin is worn away, exposing the second layer of tissue. Children do this often by "skinning" their knees. The most common cut is a clean slice called a(n) (4) _____ and is treated with stitches in extreme measures or bandaging in light wounds. The last wound discussed here is the worst. It happens when the entire limb is torn away. The (5) _____ requires bringing the lost part to the hospital for reconnection.

6. Using Figure 22-2, fill in the pulse locations being used by rescuers in each frame.

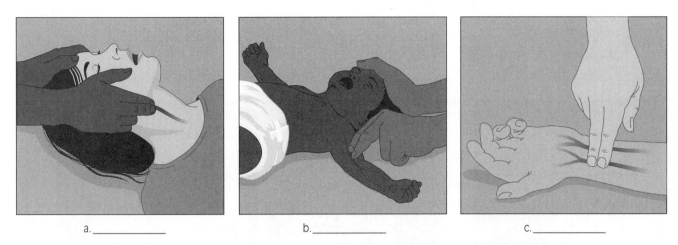

a._____    b._____    c._____

**Figure 22-2**  Locating pulses.

7. Place the correct condition that affects the pupil of the eye under the examples shown in Figure 22-3 using the following terms.

dilated

unequal

constricted

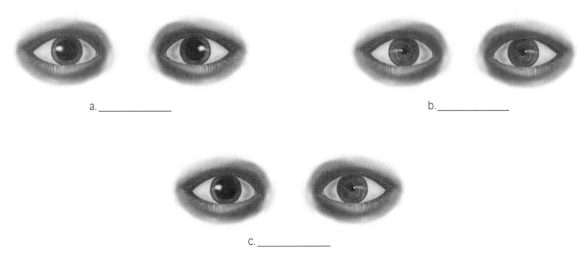

a._____    b._____

c._____

**Figure 22-3**  Examples of pupil sizes.

8. Figure 22-4 shows three different categories of burns. Place the correct burn category with each of the photos in this figure using the following terms.

superficial
partial thickness
full thickness

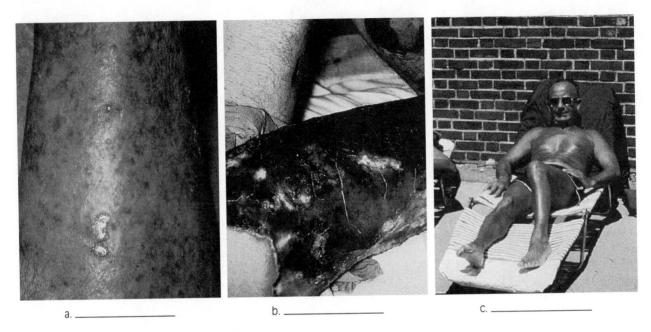

a. _____          b. _____          c. _____

**Figure 22-4**  Identify the category of burns.

# CHAPTER 23

# FIREFIGHTER SURVIVAL

# QUESTIONS

## True or False

1. Firefighter survival is pure instinct, and training will only augment that sense.

    a. true     b. false

2. If firefighters need their PPE to protect them, then other tactics, such as ventilation or proper stream placement, have failed.

    a. true     b. false

3. Most departments ask that all PPE be donned before response in order to ensure that nothing is left to chance.

    a. true     b. false

4. Relying on a company officer as the only accountability system is perhaps the best one available, since it involves personnel accountability.

    a. true     b. false

5. Freelancing is a practice that can get one injured.

    a. true     b. false

6. It is said that a common connection in injury investigation is that many injured firefighters were involved in work that they were not familiar with.

    a. true     b. false

7. One way of eliminating freelancing is to break up the teams on the fireground and allow experience to guide action.

    a. true     b. false

8. When a team is reporting updates on progress it is usually good practice to also relay hazards or conditions that may be important to the mission.

    a. true     b. false

9. If a team member is given an order and is on the way to carry out the order, and another higher ranking person gives that person another order, the person should complete the new order quickly and then the original order without delay.

    a. true     b. false

10. In rehabilitation it is recommended that firefighters remain standing so that they do not stiffen up before going back to strenuous work.

    a. true     b. false

11. It is a good practice to drink sugared beverages during rehabilitation in order to recapture energy lost in the work.

    a. true     b. false

12. The 30/30/40 plan relates to the balance of protein, fat, and carbohydrates firefighters should consume for quick energy.

    a. true     b. false

13. Rapid escape means just that. Firefighters should leave the area immediately, leaving hoselines and heavy tools behind.

    a. true    b. false

14. Most firefighters believe that being lost in an occupancy is not a true emergency, since they will eventually find a way out if they remain calm and collected.

    a. true    b. false

15. The CISD must be considered whenever anything traumatic happens on an incident that involves a firefighter's work.

    a. true    b. false

## Multiple Choice

1. Why is it important to keep turnout clothing dry?
   a. wet clothing is much heavier
   b. wet clothing may enhance electrical shock
   c. wet clothing fatigues firefighters quicker
   d. wet clothing may lead to steam burns

2. Accountability systems in use today fall into all but which of the following categories?
   a. passport
   b. triage
   c. tag
   d. company officer

3. Fitness for duty usually includes which of the following? (choose two)
   a. proper attire
   b. physical fitness
   c. mental fitness
   d. proper equipment

4. According to the text, countless investigative reports cite which of the following as a significant contributing event to the death or injury of a number of firefighters?
   a. freelancing
   b. lack of water pressure
   c. poor upkeep of equipment
   d. booby traps

5. What is the term used when the benefits a task will bring are outweighed by the hazards that task will face?
   a. benefit analysis
   b. risk/benefit
   c. sacrifice
   d. high-risk venture

6. What is the most effective way of reducing stress and overexertion on the fireground?

   a. defensive posture

   b. fog attack

   c. rehabilitation

   d. freelancing

7. Which of the following are considered key elements of rehabilitation? (choose two)

   a. rest

   b. stretching

   c. sleeping

   d. hydration

8. According to the text, firefighters should drink about how much water an hour during periods of work?

   a. 8 ounces

   b. 1 gallon

   c. 12 ounces

   d. 1 quart

9. In basic terms, maximizing energy from the human cell takes a balance of all but which of the following?

   a. carbohydrates

   b. oxygen

   c. water

   d. insulin

10. What are the two most common situations that will lead to the need for rapid escape? (choose two)

    a. water damage

    b. rapid fire spread

    c. pending water drop

    d. building collapse

11. What are the two elements that a firefighter evacuation signal must employ? (choose two)

    a. PPE

    b. teamwork

    c. air horns or sirens

    d. radios

12. What most usually will activate a personnel accountability report?

    a. payday

    b. operational change

    c. water delays

    d. air unit arrival

13. Common post-incident injuries include all but which of the following?
    a. strains
    b. being struck by objects
    c. sprains
    d. burns

14. What is the term given for the system shown in Figure 23-1, in which one person on the fireground identifies each team member?
    a. accountability
    b. positioning
    c. identification
    d. command

**Figure 23-1**  Fire officer collecting name pieces.

15. Performing a personal size-up involves evaluating all of the following *except:*
    a. hazardous energy
    b. smoke conditions
    c. officer competence
    d. escape routes

16. Which of the following are examples of active cooling? (choose two)
    a. forearm submersion
    b. resting in shade
    c. standing near fans
    d. use of misting fans

17. What foods offer the best approach to nourishment at an incident scene?
    a. low-fat carbohydrates
    b. 40/30/30 balance of carbohydrate, protein, and fat
    c. rapid-energy foods such as nuts and bananas
    d. slow-release foods such as pasta and potatoes

# HAZARDOUS MATERIALS: LAWS, REGULATIONS, AND STANDARDS

# QUESTIONS

## True or False

1. Many times at HAZMAT incidents the actions that the first-in firefighters take in the first five minutes determine how the next five hours/days/months will go.

   a. true      b. false

2. The oldest present-day HAZMAT team in the United States is over sixty years old.

   a. true      b. false

3. It is a fact that almost every substance is hazardous to human health if used improperly or if it escapes its container.

   a. true      b. false

4. OSHA and EPA regulations have the weight of law but are not congressional acts.

   a. true      b. false

5. At this time, the only people in the fire service required by law to have physicals are HAZMAT team members.

   a. true      b. false

## Multiple Choice

1. According to the text, a common device that was used strictly by HAZMAT teams, but is commonly used by engine, truck, and medic units today, is which of the following?

   a. salvage covers

   b. wading pools

   c. SCBA

   d. air monitoring devices

2. The most common definition of a hazardous material is:

   a. a substance that runs, raises, and poisons

   b. a material that is poisonous to all that it comes into contact with

   c. a material that escapes its container and hurts or harms anything it touches

   d. a substance that kills by vapor, liquid, or solid

3. If a potentially harmful chemical is released into the environment, it will specifically involve which agency?

   a. OSHA

   b. EPA

   c. NFPA

   d. NFA

4. If a chemical is a risk to employees in the workplace, it specifically comes under which regulatory agency?

   a. OSHA

   b. EPA

   c. NFPA

   d. NFA

5. What document is primarily responsible for making sure that local resources are adequate to handle a chemical release in the community?

   a. municipal bylaws

   b. local emergency response plan

   c. governmental charter

   d. Material Safety Data Sheets

6. Many chemicals must be reported by law to local authorities when stored at a facility. Some chemicals are so strong that they have separate reporting requirements and have lower reporting thresholds. What are these stronger chemicals called?

   a. high dose chemicals

   b. classified poisons

   c. extremely hazardous substances

   d. dangerous quantity chemicals

7. OSHA has designated how many levels of HAZMAT training for firefighters?

   a. three

   b. four

   c. five

   d. eight

8. Of the levels of HAZMAT training listed below, which of the following is the correct order of complexity from least to most complex?

   a. incident commander, operations, awareness

   b. awareness, incident commander, operations

   c. operations, incident commander, awareness

   d. awareness, operations, incident commander

9. The NFPA has set a number of standards for different fields of fire service endeavor. How many NFPA standards are specific to HAZMAT response and training?

   a. one

   b. two

   c. three

   d. four

10. What term best describes not following the standard of care or an accepted practice?

    a. legal obligation

    b. negligence

    c. consequence

    d. liability

11. The term used to identify the actions of the chemical shown reacting to air exposure in Figure 24-1 is which of the following?
    a. miscible
    b. soluble
    c. pyrophoric
    d. hypergolic

**Figure 24-1** The material shown here is an example of one that, when it escapes its container and comes into contact with the air, ignites.

12. How long must an employer keep any medical records related to an employee's chemical exposures?
    a. 5 years
    b. 5 years after the employee's retirement
    c. 30 years after the employee's retirement
    d. not required

13. In general, closed containers of hazardous materials present what type or level of risk?
    a. no risk at all
    b. the chemical must escape the container to create a risk
    c. always presents a risk
    d. may present a risk of flammability

14. Which level of training allows for offensive mitigation of a chemical release?
    a. incident commander
    b. Hazardous Materials Awareness
    c. Hazardous Materials Operations
    d. Hazardous Materials Technician

## Fill-in-the-Blanks

a. NFPA
b. standards
c. laws
d. SARA
e. regulations

Legislative acts that are passed by Congress are called (1) _____ , and firefighters must make sure that they follow them in their work. On the other hand, government agencies like OSHA and the Environmental Protection Agency (EPA) set up (2) _____ that have the weight of law but are not sent to Congress for passage. The SARA Act is an example of this. On a slightly lower level of law are (3) _____ , which are developed by nongovernmental agencies in a consensus fashion. One such agency is the (4) _____ , which has been influential on a national front for years in many areas beyond hazardous materials. The first law that regulated how fire departments respond to HAZMAT emergencies was the Superfund Amendment or (5) _____ , which was passed in 1986 for the protection of emergency responders and the community.

# HAZARDOUS MATERIALS: RECOGNITION AND IDENTIFICATION

*Photo courtesy of Baltimore County Fire Department*

# QUESTIONS

## Matching

Match the correct term with the definitions provided.

    a. gas
    b. solid
    c. ingestion
    d. inhalation
    e. liquid

1. The _____ hazardous material is the easiest to handle.

2. The next easiest hazardous material to handle is usually _____ .

3. The hardest hazardous material to contain and deal with is _____ .

4. If a solid hazardous material poisoned a person, the route of poisoning would most probably be _____ .

5. HAZMAT gases are usually poisonous by _____ .

6. Using the following terms, match the types of tanks to the truck photos in Figure 25-1.

    pressurized gas
    nonpressurized liquid
    corrosives

a._____

b._____

c._____

**Figure 25-1** Match the trucks to the terms listed in question 6.

# True or False

1. Knowing the chemical name is not nearly as important as knowing its classification.
   a. true     b. false

2. The DOT system attempts to provide general grouping categories for hazardous materials.
   a. true     b. false

3. Combustible liquids are those with a flash point between 0°F and 100°F.
   a. true     b. false

4. It is suspected that 50 percent to 60 percent of the trucks traveling the highways are not properly placarded.
   a. true     b. false

5. The NFPA 704 system of placarding is best used to identify specific chemicals for hazards.
   a. true     b. false

6. A good way to tell if a pipeline crosses under a roadway is to look for a sign, which is required.
   a. true     b. false

7. If pipelines should rupture, most can be shut down immediately because computer-controlled gates are located every mile per regulations.
   a. true     b. false

8. A good way to tell if many chemicals are hazardous is to check the fortification of their containers.
   a. true     b. false

9. The typical 55-gallon drum can weigh over 400 lb when full.
   a. true     b. false

10. Pipelines can originate at a bulk storage facility and then run nonstop through several states.
    a. true     b. false

11. The most common tanker truck on the road today is usually a gasoline truck.
    a. true     b. false

12. The gasoline truck is set to keep a minimum pressure of 112 psi in the tank at all times when in service.
    a. true     b. false

13. One problem with insulated tanker trucks is that leaks may travel inside and exit in a different area entirely.
    a. true     b. false

14. Pressurized tanker trucks are usually painted white in order to be seen easily at night.
    a. true     b. false

15. It is a common situation that rail shipments are more hazardous because of the greater quantities carried.
    a. true     b. false

# Multiple Choice

1. The Emergency Planning and Community Right to Know Act is a result of:
   a. Chemtrec
   b. an accident in India
   c. DOT
   d. SARA legislation

2. What is the main regulatory agency in the United States for air, water, roadways, and pipelines?
   a. NFPA
   b. Chemtrec
   c. DOT
   d. SARA

3. The DOT has established a system of how many hazard classes?
   a. three
   b. six
   c. nine
   d. twelve

4. Placards are placed on how many sides of a vehicle?
   a. one
   b. two
   c. three
   d. four

5. What part of the DOT system requires placarding at 1,001 lb and above?
   a. guidebook
   b. Table 2
   c. code of federal regulations
   d. Chart 9

6. Poisonous gases are separated into zones. What zone is the most hazardous?
   a. A
   b. B
   c. C
   d. D

7. Poisons are designated Class 6 and then divided into divisions. Which division is the most hazardous?
   a. 1
   b. 2
   c. 3
   d. 4

8. In what section of the 704 diamond would one look to find if the material was radioactive?
   a. health
   b. reactivity
   c. flammability
   d. special hazard

9. If one were to find a "w" with a slash through it on the 704 placard and a numeral 1 next to it, that would usually mean:
   a. explosion can result
   b. foam is required
   c. contents may react with water
   d. no such placard

10. According to the text, which type of drum is commonly used to store corrosives such as hydrochloric acid?
    a. fiberboard
    b. plastic
    c. steel
    d. aluminum

11. Cylinder-shaped containers share a common connection. That connection is which of the following?
    a. pressurization
    b. weight
    c. liquids are present
    d. container weakness

12. Which of the following tanker trucks are known to have the most accidents on the roads in the United States?
    a. corrosives
    b. pressurized gases
    c. heated fluids
    d. gasoline

13. The tanker trucks that have rounded ends and smooth exteriors commonly carry which of the following commodities?
    a. gasoline
    b. corrosives
    c. waste material
    d. gases

14. In a pressurized gas truck it is common practice to fill it no more than:
    a. 99 percent
    b. 90 percent
    c. 80 percent
    d. 70 percent

15. Upright storage tanks that are constructed of steel and typically have a sloped or cone-shaped roof to shed rainwater are commonly called:

    a. ordinary tanks

    b. floating roof tanks

    c. lifter roof tanks

    d. pressure relief tanks

16. What type of roof on a standing tank is designed to float on the liquid within?

    a. dome roof

    b. flat roof

    c. external floating roof

    d. none of the above

17. Gases with a vapor density of 3 will:

    a. rise and dissipate

    b. rise and stay in a cloud

    c. mix with water

    d. drop to the earth

18. A liquid with a specific gravity of 3 would most likely:

    a. rise and vaporize

    b. rise and float

    c. mix with water

    d. sink in the water

19. If a material is water soluble it is said to:

    a. rise and vaporize

    b. rise and float

    c. mix with water

    d. sink in water

20. What is the term used when a material gives off vapors that will ignite but quickly go out, not sustaining combustion?

    a. vaporization

    b. exothermic reaction

    c. volatility

    d. flash point

21. Figure 25-2 shows one of the most common HAZMAT incident potentials. Which one of the following types of incident potentials is depicted here?

    a. corrosive spill

    b. propane leak

    c. nitric acid spill

    d. gasoline spill

**Figure 25-2**   If a community has a road, the potential for a hazardous materials incident exists.

22. Identify the placard style shown in Figure 25-3.

    a. NFPA 704

    b. NFPA 1910

    c. DOT 909

    d. DOT ID

**Figure 25-3**   Placard system.

23. What is the biggest risk when tanks, trucks, and tank cars are involved in fires?
    a. spills
    b. leaks
    c. smoke
    d. BLEVE

24. When does the BLEVE time bomb clock start to tick?
    a. from the 911 call
    b. upon the first application of heat to the tank
    c. only in the winter
    d. when the first engine company arrives

25. Which should cause an immediate withdrawal of firefighters from a propane tank fire?
    a. rising sound from the venting relief valves
    b. discoloration of the tank
    c. both a and b
    d. none of the above

## Fill-in-the-Blanks

    a. red
    b. 4
    c. health
    d. 0 - 4
    e. yellow

The NFPA 704 diamond is a common system of identifying hazards in occupancies. The blue section denotes
(1) _____ hazard, and the (2) _____ designates fire hazard. The
(3) _____ color is set up to identify reactivity, and the numerals inside these sections can run
(4) _____ , with (5) _____ being the most hazardous.

# HAZARDOUS MATERIALS: INFORMATION RESOURCES

# QUESTIONS

## True or False

1. When shipping hazardous chemicals, it is common practice for both the shipper and the facility to maintain certain documents to assist first responders.

   a. true      b. false

2. One shortcoming of the DOT ERG is that it fails to provide evacuation distances, which many other publications do.

   a. true      b. false

3. The DOT ERG is one of the more complete reference books, since it contains up to 4,000 chemicals and as many guide pages.

   a. true      b. false

4. Highway shipping papers are carried by the driver or are at least within arm's reach during transportation.

   a. true      b. false

5. Chemtrec was the largest information source in the United States until its demise a few years ago.

   a. true      b. false

6. If a material has the ability to polymerize, it will be noted in the DOT ERG.

   a. true      b. false

7. If the DOT ERG recommends that vapors be knocked down with a hoseline, this action should always be taken.

   a. true      b. false

## Multiple Choice

1. The DOT's Emergency Response Guidebook is better known as which of the following?

   a. quick reference
   b. Cameo
   c. the orange book
   d. firefighter awareness

2. The DOT book assigns identifying numbers to each chemical in what fashion?

   a. alphabetically
   b. by four-digit numbers
   c. by hazard rating
   d. by fire or explosion category

3. Which of the following is true about the DOT Guidebook? (choose two)

   a. more value to the initial responder
   b. less value to the initial responder
   c. more value to the HAZMAT technician
   d. less value to the HAZMAT technician

4. Some of the hazards associated with using water to knock down vapors include all but which one of the following?

   a. increase of spill size

   b. possible reactions

   c. creating runoff problems

   d. water conducts electricity

5. Which is true regarding most hazardous chemical releases?

   a. they travel farther in the cool of the night

   b. they travel farther in the heat of the day

   c. they are all endothermic

   d. all containers are subject to BLEVE

6. According to the text, OSHA 29 CFR 1910.1200 requires an MSDS for which of the following?

   a. anything liquid

   b. any flammable chemical

   c. all vapor-producing chemicals

   d. any HAZMAT above household quantity

7. Given the choice, according to the text, the first responder should rely more on the technical information supplied by which of the following forms?

   a. waybill

   b. DOT

   c. MSDS

   d. tech sheet

8. Which form of transportation commonly uses the bill of lading?

   a. aircraft

   b. watercraft

   c. rail

   d. highway

9. Which of the following forms of transportation commonly uses the waybill?

   a. aircraft

   b. watercraft

   c. rail

   d. highway

10. The pilots of an aircraft carrying HAZMAT must have which of the following in their possession?

    a. waybill

    b. bill of lading

    c. cargo manifest

    d. air bills

11. The highlighting used for some chemicals in the DOT pages in Figure 26-1 is used to depict which of the following?

    a. double meaning

    b. evacuation may be needed

    c. most explosive

    d. most viscous

| ID No. | Guide No. | Name of Material |
|--------|-----------|------------------|
| 2465 | 140 | Sodium dichloroisocyanurate |
| 2465 | 140 | Sodium dichloro-s-triazinetrione |
| 2466 | 143 | Potassium superoxide |
| 2467 | 140 | Sodium percarbonates |
| 2468 | 140 | Trichloroisocyanuric acid, dry |
| 2468 | 140 | Trichloro-s-triazinetrione, dry |
| 2468 | 140 | (mono)-(Trichloro)-tetra-(monopotassium dichloro)-penta-s-triazinetrione, dry |
| 2469 | 140 | Zinc bromate |
| 2470 | 152 | Phenylacetonitrile, liquid |
| 2471 | 154 | Osmium tetroxide |
| 2473 | 154 | Sodium arsanilate |
| 2474 | 157 | Thiophosgene |
| 2475 | 157 | Vanadium trichloride |
| 2477 | 131 | Methyl isothiocyanate |
| 2478 | 155 | Isocyanate solution, flammable, poisonous, n.o.s. |
| 2478 | 155 | Isocyanate solution, flammable, toxic, n.o.s. |
| 2478 | 155 | Isocyanate solutions, n.o.s. |
| 2478 | 155 | Isocyanates, flammable, poisonous, n.o.s. |
| 2478 | 155 | Isocyanates, flammable, toxic, n.o.s. |
| 2478 | 155 | Isocyanates, n.o.s. |
| 2480 | 155 | Methyl isocyanate |
| 2481 | 155 | Ethyl isocyanate |
| 2482 | 155 | n-Propyl isocyanate |
| 2483 | 155 | Isopropyl isocyanate |
| 2484 | 155 | tert-Butyl isocyanate |
| 2485 | 155 | n-Butyl isocyanate |
| 2486 | 155 | Isobutyl isocyanate |
| 2487 | 155 | Phenyl isocyanate |

**Figure 26-1** Why are some of these chemicals highlighted?

12. Given the information provided by the EPA on the use of Material Safety Data Sheets (MSDS) a responder:

    a. should always follow the advice on the MSDS

    b. never follow the advice on the MSDS

    c. use the MSDS in conjunction with other reference sources

    d. always call the trucking company

# HAZARDOUS MATERIALS: PERSONAL PROTECTIVE EQUIPMENT

# QUESTIONS

## Matching

Match the correct term with the definitions provided.

     a. ceiling levels

     b. TRACEM

     c. permeation

     d. irritant

     e. carcinogen

1. One of the most common acronyms used today for the methods of identifying the hazards of different chemical releases is _____ .

2. A(n) _____ is a term used to inform the public that the item is a cancer-producing material.

3. When a chemical affects the body in such a way as a corrosive material might, it is considered a(n) _____ and should be taken care of as soon as possible.

4. _____ are the highest amounts of chemical exposure a person can be exposed to before becoming affected.

5. The movement of a chemical through the fabric of protective clothing is called _____ , which is what determines that fabric's level of protection.

6. Match the following terms to the drawings in Figure 27-1, naming the proper type of chemical exposure.

     ingestion        inhalation        injection        absorption

     a. _____

     b. _____

     c. _____

     d. _____

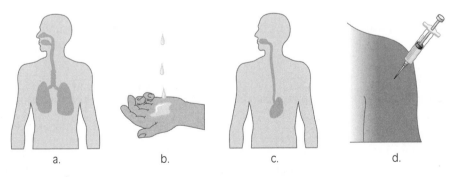

**Figure 27-1** Routes of exposure.

7. Which one of the suits in Figure 27-2 is a Level A suit? Circle one:

    a   or   b

a.

b.

**Figure 27-2**   Which suit is a Level "A" suit?

# True or False

1. Firefighter turnout gear is frequently tested for its ability to protect the wearer during a chemical spill or splash on the clothing.

    a. true      b. false

2. According to the text, the human body does well with short-duration exposures and does recover from these exposures.

    a. true      b. false

3. Of the more than 21 million chemicals on this Earth, only about twelve have been identified as known cancer-causing agents.

    a. true      b. false

4. When considering exposure, it must be remembered that the lungs are the body's largest organs and must be protected.

    a. true      b. false

5. It is a proven fact that most firefighters do a good job of decontaminating themselves prior to taking rehabilitation food and beverages.

a. true     b. false

6. The reason many firefighters are not affected by minor chemical exposure is because their increased body temperatures and respiratory rates ward off chances of contamination.

a. true     b. false

7. Donning a chemical protective suit further limits the field of vision from the normal SCBA face piece range.

a. true     b. false

8. Newly developed technology has finally created a chemical suit that will protect the firefighter from just about all chemical exposures.

a. true     b. false

9. With the recirculating effect of the newer SCBA, heat stress is no longer a concern when wearing the chemical suits.

a. true     b. false

10. Turnout gear will slow the chemical exposure down but will not prevent the eventual migration of the chemical to the skin.

a. true     b. false

11. The HAZMAT suit shown in Figure 27-3 is the ultimate protective garment used in today's fire service.

a. true     b. false

**Figure 27-3**   Ultimate protection?

# Multiple Choice

1. Which of the following items will offer the greatest amount of chemical protection in all cases of exposure?
   a. gloves
   b. turnout coat
   c. encapsulating suit
   d. SCBA

2. The study of poisons and their effect on the body is called:
   a. technology
   b. scientology
   c. toxicology
   d. oncology

3. What are the two types of chemical exposure? Choose two.
   a. habitual
   b. chronic
   c. high
   d. acute

4. Which of the following is *not* one of the primary routes of exposure to the body?
   a. absorption
   b. laceration
   c. injection
   d. respiratory

5. The route taken by a hazardous material that is most commonly associated with causing detrimental effects to human health is which of the following?
   a. absorption
   b. laceration
   c. injection
   d. respiratory

6. Complete the following equation: effect = dose + concentration + ?
   a. time
   b. effort
   c. circumstances
   d. material

7. Which of the following is the only agency that provides legally binding exposure levels of chemicals in the United States?
   a. NFPA
   b. IAFF
   c. OSHA
   d. FEMA

8. According to the text, wearing SCBA may increase survivability by how many times versus not wearing any?

    a. 50

    b. 100

    c. 1,000

    d. 10,000

9. Although there are a lot of different types of SCBA available, the most common for use with hazardous materials is:

    a. thirty minute

    b. sixty minute

    c. ninety minute

    d. rebreather

10. The four levels of protective clothing are Level A, B, C, and D. Which level is the least protective?

    a. A

    b. B

    c. C

    d. D

11. No matter the type of PPE, what is always a concern?

    a. shoe size

    b. heat stress

    c. mechanical protection

    d. heat protection

12. What is the major difference between Level B and Level C protective ensembles?

    a. color

    b. size

    c. a respirator

    d. boot length

13. What is one easy way to prevent cancer as a result of hazardous materials exposure?

    a. always wear your SCBA when in hazardous situations

    b. always wear gloves

    c. eat lettuce daily

    d. wear a Level C ensemble for chemical situations

# CHAPTER 28

# HAZARDOUS MATERIALS: PROTECTIVE ACTIONS

*Courtesy of Baltimore County Fire Department*

# QUESTIONS

## Matching

Put the following steps to decontamination into the correct order:

a. body wash            b. gross decontamination

c. SCBA removal        d. medical evaluation

e. tool drop              f. PPE removal

g. clothing removal     h. formal decontamination

i. dry off                j. return to the station

1. _____      2. _____

3. _____      4. _____

5. _____      6. _____

7. _____      8. _____

9. _____      10. _____

11. Which of the examples shown in Figure 28-1 is an example of gross decontamination?

     a or b (circle one)

a.

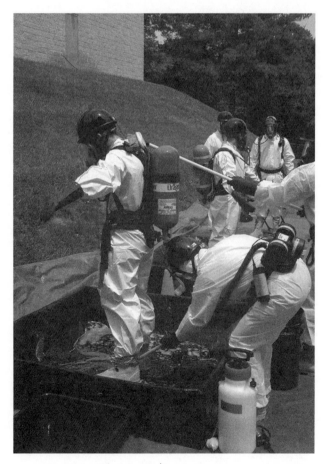

b.

**Figure 28-1** Which of these is a gross decontamination?

## True or False

1.  While each chemical spill is different, the procedures for handling them remain the same.
    a. true    b. false

2.  The OSHA HAZWOPER rules require the use of an incident management system (IMS) on HAZMAT spills.
    a. true    b. false

3.  The decision to execute a rescue is a personal one, since it may involve substantial risk to the rescuer.
    a. true    b. false

4.  Before moving the HAZMAT victim from the location of the accident, it is imperative that vitals are taken and lifesaving medical procedures are started.
    a. true    b. false

5.  In most cases, the HAZMAT team is the technical group that performs the mitigation of the incident.
    a. true    b. false

6.  A person trained to the operations level is not adequate to be the safety officer for the HAZMAT operation.
    a. true    b. false

7.  OSHA rules state that the number of entry team members to enter the hazard area shall be a minimum of two.
    a. true    b. false

8.  The DOT Guidebook has established a minimum of 50 feet as an isolation area for unknown substances.
    a. true    b. false

9.  Many HAZMAT scenes are set up into zones. PPE is not required for cold zones because there should be no chance for chemical exposure.
    a. true    b. false

10. The use of air monitors is important when setting up isolation points on the HAZMAT scene.
    a. true    b. false

11. When sheltering in place, it is a good idea to place an emergency responder at the sites with large numbers of people.
    a. true    b. false

12. The general rule for dealing with explosives containers on fire is to hit them quickly (if burning less than five minutes) with great quantities of water and then back off, leaving staffed monitors.
    a. true    b. false

13. The most common incident involving natural gas is the ruptured gas main.
    a. true    b. false

14. If a pressurized tank full of propane is exposed to fire, the responder is safe as long as the tank relief valve is open, no matter what other signs are present.
    a. true    b. false

15. A little known fact is that a small percent (5 percent or less) of gasoline is a confirmed cancer-causing agent.

a. true      b. false

16. The material commonly kept in the bottles shown in Figure 28-2 is known to be lighter than common air.

a. true      b. false

**Figure 28-2** These 20-pound cylinders, found in most homes, can create large fireballs and can explode with considerable force if involved in fire.

## Multiple Choice

1. When discussing HAZMAT response, it is said that the basic concept for first responders is that of:

a. isolation

b. exposure

c. control

d. management

2. When dealing with exposed people at a HAZMAT incident, who holds the responsibility for their movement in and out of the exposure area?

   a. incident commander

   b. safety team

   c. HAZMAT team

   d. rescue team

3. When rescuing victims, there should be no delay. Which of the following techniques is recommended by the text?

   a. mitigation release

   b. swoop and scoop

   c. move and decontamination

   d. capture of hazard

4. A system designed to identify the risk that chemicals present is called:

   a. DOT identification

   b. risk categorizing

   c. mitigation minimizing

   d. risk-based response

5. An unknown chemical can be divided into one of three risk categories. Which of the following is *not* one of those categories?

   a. fire

   b. reactive

   c. corrosive

   d. toxic

6. The major determining factor in the hazard that a chemical presents is which of the following?

   a. vapor pressure

   b. toxicity

   c. flammability

   d. reactivity

7. No matter who is performing the tasks at an incident, which of the following people is responsible for the overall actions of the incident?

   a. HAZMAT team leader

   b. safety officer

   c. operations chief

   d. incident commander

8. At a HAZMAT incident, there are two similar positions involved in safety. What are their responsibilities? (choose two)

   a. overall scene safety

   b. HAZMAT specific safety

   c. decontamination safety

   d. cool zone safety

9. It is recommended to have a backup team at expanded incidents. What level of training must be considered minimum for this team?

   a. operational

   b. specialist

   c. technician

   d. command

10. The initial isolation area at an incident later becomes which of the following as the HAZMAT team begins to set up?

   a. decontamination

   b. warm zone

   c. hot zone

   d. safety zone

11. List two items mentioned in the text that many first arriving companies forget to communicate upon arrival at a HAZMAT incident. (choose two)

   a. scene description

   b. wind direction

   c. best travel route for others

   d. amount of material involved

12. Which zone is the best location for the Command Post?

   a. hot

   b. cold

   c. warm

   d. isolation

13. Which of the following best describes sheltering in place?

   a. shut down HVAC systems

   b. remove all persons on foot

   c. provide evacuation locations

   d. stay indoors and close all openings

14. Which of the following describes the two types of release categories of hazardous materials? (choose two)

   a. blown valve

   b. pressure release of gases

   c. breach of a container

   d. release within a containment system

15. Where would a bomb squad be best utilized?

   a. explosives incidents

   b. reactive incidents

   c. terrorism emergencies

   d. government buildings

16. Name the two most commonly released flammable gases. (choose two)

    a. gasoline

    b. natural gas

    c. butane

    d. propane

17. If a gas main in the street should rupture, what would be the best action for responding units?

    a. shut off, cool, protect

    b. evacuate, extinguish, bury

    c. isolate, protect, stand by

    d. extinguish, evacuate, cap

18. What is the leading chemical involved when it comes to chemical accident fatalities?

    a. propane

    b. acetylene

    c. acid

    d. gasoline

19. Which of the following is very dangerous when spilled on asphalt, since it becomes shock sensitive?

    a. butane

    b. liquid oxygen

    c. nitrogen

    d. carbon dioxide

20. Which one of the following hazardous materials is commonly used in smoke detectors, in ground-imaging equipment, and in the medical community?

    a. radioactive material

    b. carcinogenic material

    c. toxic chemicals

    d. pressurized materials

21. What is the primary hazard with magnesium as shown in Figure 28-3?

    a. poisonous

    b. explosive

    c. water soluble

    d. water reactive

**Figure 28-3** Burning magnesium.

22. A citizen brings in a pipe bomb to the fire station. Firefighters should
    a. handle the device
    b. remove any fuse
    c. bring it in and call the bomb squad
    d. leave it where it is and call the bomb squad

23. A car that is running on CNG or LNG generally:
    a. has a sticker on it indicating that fact
    b. gets poorer fuel mileage than a gasoline car
    c. only runs during the day
    d. presents no risk to responders

24. What are the two major categories of radiation?
    a. alpha and X ray
    b. X ray and gamma
    c. neutron and sunlight
    d. ionizing and non-ionizing

# Fill-in-the-Blanks

1. Name the three zones at the HAZMAT spill shown in Figure 28-4.


_____ , _____ , _____


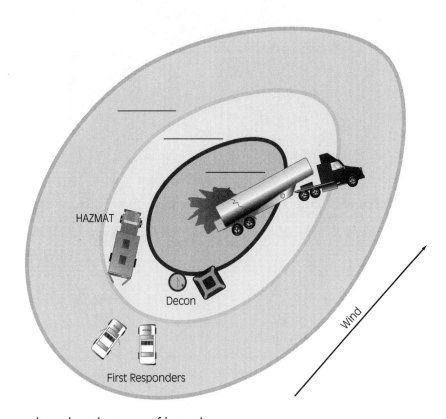

**Figure 28-4**  Zones are based on the types of hazards.

2. Describe the process shown in Figure 28-5.

a. _____

b. _____

c. _____

d. _____

e. _____

a.

b.

c.

d.

e.

**Figure 28-5**  Describe the process shown here.

# CHAPTER 29

# PRODUCT CONTROL AND AIR MONITORING

# QUESTIONS

## True or False

1. When considering product control, it is standard practice to reduce the surface area of the material, which will provide a direct reduction in the danger to the responders.

    a. true     b. false

2. Most spilled hazardous materials will sink, requiring an overflow dam with a 4-inch pipe to draw off the material.

    a. true     b. false

3. One method of HAZMAT reduction is to build a retention container. This is most often done by damming up storm drains.

    a. true     b. false

4. Since both natural gas and propane are water soluble, using water spray will neutralize their hazards.

    a. true     b. false

5. All HAZMAT gas monitors have a lag time before actual reading is possible.

    a. true     b. false

6. Absorbent pads such as those shown in Figure 29-1 are used to absorb water so the oils can be picked up easily.

    a. true     b. false

**Figure 29-1** Using absorbent pads.

# Multiple Choice

1. What type of operation would absorption, diking, and retention be?
   a. offensive
   b. aggressive
   c. defensive
   d. passive

2. When constructing a dam or dike, it is best to follow certain rules. Where should the first dam be constructed when setting up multiple dams?
   a. as close to the incident as possible
   b. as far from the incident as possible
   c. in the spill itself to slow the flow
   d. at the leading edge of the spill

3. Name the two basic types of dams created in HAZMAT incidents. (choose two)
   a. back wall
   b. underflow
   c. overflow
   d. retainer

4. What will make the use of water vapor or water fog most effective on hazardous materials?
   a. chemical must be water soluble
   b. wind conditions
   c. topography must be right
   d. the proper nozzle diameter

5. Using foam to suppress vapors is generally limited to which of the following hazards?
   a. flammable gases
   b. water-soluble solids
   c. oxidizers
   d. flammable liquids

6. Most tanker trucks on today's highways have two emergency shutoff valves in case of main valve failure. Where are these emergency valves usually located? (choose two)
   a. cab of vehicle
   b. behind cab of vehicle on tank
   c. center rear of tank
   d. near the main control valves

7. What is the term used for testing a gas monitor for a known gas and allowing it to sound an alarm, then removing the gas quickly?
   a. spurt test
   b. high flow test
   c. quantitative test
   d. bump test

8. According to the text, most of the new flammable gas indicators are used to measure which of the following?

   a. lower explosive limits

   b. upper explosive limits

   c. toxicity

   d. only programmed materials

9. As a direct result of confined space regulations, the most common gas monitors today typically measure how many gases?

   a. one

   b. two

   c. four

   d. a dozen

10. A common gas detector in the residential setting measures which of the following hazardous materials?

    a. carbon dioxide

    b. oxygen

    c. nitrogen

    d. carbon monoxide

11. What is the most common LEL sensor used today?

    a. catalytic bead

    b. metal oxide

    c. metal dioxide

    d. infrared bridge

12. The LEL for methane is 5%. If a LEL sensor calibrated for methane is reading 50% how much methane is present in the room?

    a. 5%

    b. 2.5%

    c. 100% by volume

    d. none

13. What meter is used to identify potentially toxic risks?

    a. LEL sensor

    b. oxygen sensor

    c. PID

    d. infrared sensor

## Fill-in-the-Blanks

1. Figure 29-2 shows two types of dams commonly used by fire department HAZMAT teams to contain hazardous materials. Name each type shown.

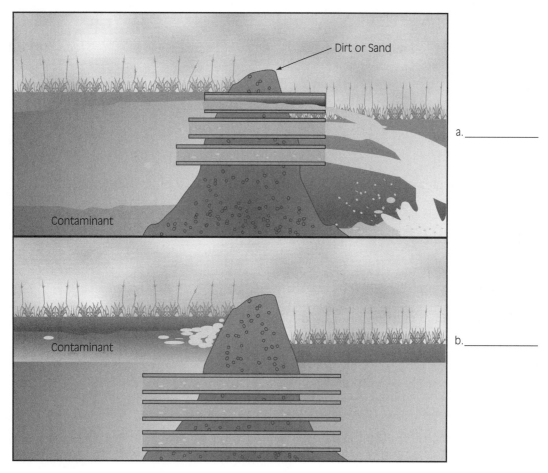

**Figure 29-2** Containment dams.

# CHAPTER 30

# TERRORISM AWARENESS

# QUESTIONS

## Matching

Match the correct term with the definitions provided.

     a.  hydrogen cyanide

     b.  tear gas

     c.  anthrax

     d.  sarin

     e.  Molotov cocktail

1.  Nerve agents include agents such as _____ .

2.  The _____ is an example of an incendiary agent.

3.  _____ is a well-known blood agent, which as a gas affects the body's use of oxygen.

4.  Irritants such as _____ are very common terrorism agents.

5.  _____ is a good example of a biological agent and toxin.

## True or False

1.  The FBI defines terrorism as a passive act or dangerous activity that furthers political gain.
    a.  true       b.  false

2.  It is considered a good idea for first responders to know the location of target facilities in their jurisdiction.
    a.  true       b.  false

3.  It is common for first responders to regard all explosions as possible acts of terrorism, but they must discontinue this practice because it jeopardizes their awareness levels.
    a.  true       b.  false

4.  In most cases, the national media does not know or follow local protocols and will require information at terrorism scenes.
    a.  true       b.  false

5.  Because of secondary device potential it is good practice to stage second-in units away from the immediate scene until an all clear is given.
    a.  true       b.  false

## Multiple Choice

1.  The best defense against terrorism is which of the following?

    a.  counterattack

    b.  education

    c.  HAZMAT training

    d.  security

2. The types of terrorism are divided into two distinct areas. They are: (choose two)
   a. foreign based
   b. high hazard
   c. progressive
   d. domestic

3. The most common device used in terrorism today is which of the following items?
   a. Molotov cocktail
   b. high explosive
   c. pipe bomb
   d. gas grenade

4. No matter what type of device a firefighter locates, the first thing to do is:
   a. ask for ambulances
   b. call for bomb squad
   c. alert the media
   d. isolate the area

5. If terrorism is suspected in an incident, who will be in charge as the situation stabilizes?
   a. bomb squad
   b. FBI
   c. highway patrol
   d. military

6. Companies arriving at an explosion (suspected terrorism incident) should avoid: (choose two)
   a. dead-end streets
   b. mailboxes
   c. grocery stores
   d. intersections

7. The most common targets are all but which of the following?
   a. schools
   b. malls
   c. theaters
   d. police stations

8. According to the text, the government of the United States has set up which of the following to counter the effects of terrorism in the United States and abroad?
   a. twenty-five USAR teams
   b. updated logistics
   c. media instruction
   d. counterterrorism teams

9. Most terrorist methods utilize toxic agents that have which of the following in common?
   a. poison through ingestion
   b. poison through inhalation
   c. poison through injection
   d. poison through contact

10. Which of the following does not fit in the "quick-in/quick-out" approach to an incident?
   a. do not treat victims; remove them
   b. keep in mind the terrorist may be among the injured
   c. secure the scene before beginning any triage
   d. watch for secondary devices

11. In this day and age, any *suspicious* package should be suspected of containing what?
   a. anthrax
   b. ricin
   c. sarin
   d. explosives

12. What is the most likely illegal lab?
   a. drug
   b. explosive
   c. chemical agent
   d. biological agent

13. A HAZMAT crime is one that:
   a. uses guns as weapons
   b. involves the stealing of chemicals
   c. is using chemicals as the weapon
   d. is drug related

# CHAPTER ONE ANSWERS

## Matching

1. b
2. e
3. c
4. a
5. d

## True or False

1. b
2. b
3. a
4. b
5. a

## Multiple Choice

1. b
2. c
3. d
4. a
5. c
6. b
7. c
8. a
9. b
10. c

# CHAPTER TWO ANSWERS

## Matching

1. a
2. e
3. f
4. i
5. k
6. m
7. n
8. d

9. g

10. o

11. l

12. j

13. b

14. c

15. h

16. p

17. Answers for Figure 2-1

    1. b

    2. d

    3. e

    4. a

    5. c

## True or False

1. a

2. a

3. a

4. b

5. b

6. a

7. b

8. a

9. b

10. a

## Multiple Choice

1. d

2. a

3. b

4. c

5. d

6. a, d–any order is acceptable

7. b

8. b, c–any order is acceptable

9. c

10. a

11. d

12. b

13. b

14. a

15. c

16. d

17. b

18. a

19. c

20. d

21. c

22. a

23. d

## Fill-in-the-Blanks

1. b

2. e

3. g

4. j

5. h

6. d

7. i

8. f

9. c

10. a

11. k

12. a. rescue specialist or technical rescue

    b. paramedic or emergency medical services

    c. hazardous materials team member

# CHAPTER THREE ANSWERS

## Matching

1. c

2. g

3. f

4. b

5. e

6. j

7. i

8. h

9. a

10. d

## True or False

1. b

2. a

3. b

4. a

5. b

6. b

7. a

8. b

9. a

10. b

11. a

12. a

13. b

14. a

15. b

16. a

# Multiple Choice

1. d

2. a

3. b

4. c

5. a

6. c

7. d

8. a

9. d

10. c

11. c

12. a

13. c

14. d

15. c

16. b

17. a

18. c

19. a, c–any order is acceptable

20. b

21. c

# Fill-in-the-Blanks

1. b

2. e

3. d

4. a

5. c

# CHAPTER FOUR ANSWERS

## Matching

1. c
2. a
3. b
4. e
5. d
6. f
7. 1. D
   2. B
   3. A
   4. C
8.

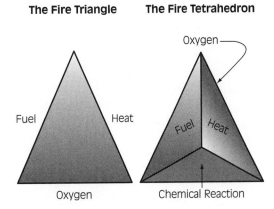

**Figure 4-2**  Fire triangle and tetrahedron.

## True or False

1. a
2. b
3. a
4. b
5. a
6. b
7. b
8. a
9. a
10. b
11. a

# Multiple Choice

1. b
2. c
3. d
4. a
5. b
6. c
7. d
8. a
9. b
10. b
11. c
12. c, d–any order is acceptable
13. c
14. d
15. d
16. b
17. b, d–any order is acceptable
18. c
19. a
20. c
21. d
22. a
23. d

# Fill-in-the-Blanks

1. d
2. b
3. e
4. a
5. c
6. g
7. f
8. h
9. friction
10. a. greater than 1
    b. less than 1
11. a. ignition
    b. growth stage
    c. fully developed stage
    d. decay stage

# CHAPTER FIVE ANSWERS

## Matching

1. d
2. a
3. e
4. c
5. b
6. a. g
   b. h $\Big\}$ any order is acceptable
   c. f

## True or False

1. a
2. b
3. b
4. a
5. a
6. b
7. a
8. b
9. b
10. a

## Multiple Choice

1. b
2. c
3. d
4. a
5. b
6. c
7. d
8. a
9. d
10. d
11. d
12. b
13. a
14. d
15. b, d–any order is acceptable
16. c
17. b
18. c
19. d

# CHAPTER SIX ANSWERS

## True or False

1. a
2. a
3. a
4. b
5. b
6. a
7. b
8. b

## Multiple Choice

1. d
2. a, c–any order is acceptable
3. b
4. a
5. a
6. c
7. a, b–any order is acceptable
8. d
9. b
10. c
11. d
12. d
13. b
14. a
15. a
16. c
17. a
18. c
19. d

## Fill-in-the-Blanks

1.–5. Five answers are needed; all eight choices are correct.

NFPA
ANSI
OSHA
IAFF
CDC
EPA
NIOSH
ASTM
⎬ any order is acceptable

6. team or buddy check

# CHAPTER SEVEN ANSWERS

## Matching

1. c
2. b
3. e
4. a
5. d
6. f

## True or False

1. b
2. a
3. a
4. b
5. a
6. b
7. b
8. b
9. a
10. a
11. b
12. a
13. a

## Multiple Choice

1. d
2. a, d–any order is acceptable
3. c
4. a
5. c
6. b
7. a
8. a, b–any order is acceptable
9. a, d–any order is acceptable
10. d
11. b
12. a
13. d
14. a
15. d
16. d

## Fill-in-the-Blanks

1. e
2. f
3. d
4. g
5. c
6. h
7. b
8. i
9. a
10. j
11. 1. backpack/harness
    2. cylinder
    3. regulator
    4. face piece
12. Agricultural occupancies will usually have chemicals and pesticides on hand.
13. The high pressure line may still be pressurized.

# CHAPTER EIGHT ANSWERS

## Matching

1. d
2. b
3. c
4. a
5. e
6. 1. a
   2. f
   3. d
   4. g
   5. b
   6. c
   7. e
7. 1. b
   2. c
   3. d
   4. a

## True or False

1. b
2. a
3. b

4. b

5. b

6. a

7. a

8. a

9. b

10. a

11. a

12. a

## Multiple Choice

1. c, d–any order is acceptable

2. d

3. a

4. b, c–any order is acceptable

5. c, d–any order is acceptable

6. b

7. c

8. a, d–any order is acceptable

9. c

10. b

11. a, d–any order is acceptable

12. c

13. a

14. d

15. d

16. b

17. c

18. c

19. d

# CHAPTER NINE ANSWERS

## Matching

1. b

2. d

3. a

4. e

5. c

6. 1. b

   2. a

7. From left to right: c, a, b

8. a. gravity-fed

   b. direct pump

   c. combination gravity-pumped system

# True or False

1. a
2. b
3. b
4. b
5. a
6. b
7. a
8. b
9. b
10. b
11. a
12. a
13. b
14. a
15. a

# Multiple Choice

1. d
2. d
3. c
4. b
5. a
6. a
7. c
8. a, b–any order is acceptable
9. c
10. d
11. b, c–any order is acceptable
12. d
13. a, c–any order is acceptable
14. c
15. c
16. a
17. c
18. a
19. c
20. a
21. b
22. d

## Fill-in-the-Blanks

1. c
2. e
3. b
4. d
5. a

Figure 9-4

6. f

# CHAPTER TEN ANSWERS

## Matching

1. c
2. e
3. a
4. d
5. b
6. a. straight
   b. single donut
   c. twin donut

## True or False

1. a
2. a
3. b
4. b
5. a
6. b
7. a
8. b
9. b
10. a
11. a
12. b
13. b
14. b
15. b

## Multiple Choice

1. a
2. b
3. c

4. d

5. c

6. d

7. a

8. b

9. c

10. d

11. c, d–any order is acceptable

12. b

13. b, c–any order is acceptable

14. c

15. a

16. a

17. b

18. d

## Fill-in-the-Blanks

1. e  } any order is acceptable
2. j

3. g

4. a

5. h

6. b

7. i  } any order is acceptable
8. c

9. f

10. d

11. a. straight load

    b. flat load

    c. horseshoe load

12. c. The firefighter starts at one end of the hose and begins to load hose over the shoulder.

    a. A fold is created, and the hose is laid on itself back to the front.

    b. The firefighter continues to walk forward folding the hose at the waist until finished.

13. Advancing an accordion load. (a) A nozzle or other appliance is attached to the end of the hose, and a desired amount is selected. (b) The firefighter pulls the amount of hose halfway out of the bed and places it on the shoulder with the nozzle or appliance on the bottom nearest the body. (c) The firefighter reaches up and pulls a number of folds from the bed and drops them on the ground at the rear of the engine, since the firefighter will drag that hose straight when walking forward. (d) When the point is reached where the hose on the firefighter's shoulder begins to pull off, the firefighter will release it a fold at a time, flaking it off behind while walking until the nozzle or appliance is all that is left on the shoulder.

14. Progressive hose lay. It is used most often in the wildland environment.

# CHAPTER ELEVEN ANSWERS
## Matching

1. b
2. c
3. a
4. e
5. d
6. a. break-apart playpipe
   b. rotating with playpipe
   c. built-in lever with pistol grip
   d. built-in lever
7. a. bank-in technique
   b. bank-back technique
   c. raindown technique
8.

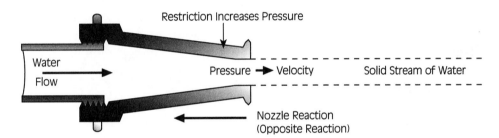

**Figure 11-3**  Nozzle reaction.

## True or False

1. a
2. b
3. b
4. a
5. a

## Multiple Choice

1. c, d–any order is acceptable
2. a, c–any order is acceptable
3. a
4. a
5. a
6. a

7. c

8. d

9. a

10. a

11. c

12. b

13. b, d–any order is acceptable

14. c

15. d

16. a

17. c

18. b

19. c

20. b

21. c

22. b

23. a

## Fill-in-the-Blanks

1. e  } any order is acceptable
2. b

3. d

4. c

5. a

# CHAPTER TWELVE ANSWERS

## Matching

1. b

2. d

3. e

4. a

5. c

6. g

7. j

8. i

9. f

10. h

11. b

12.

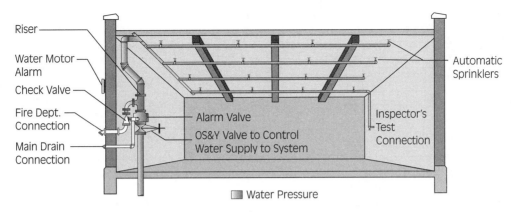

Riser
Water Motor Alarm
Check Valve
Fire Dept. Connection
Main Drain Connection
Alarm Valve
OS&Y Valve to Control Water Supply to System
Automatic Sprinklers
Inspector's Test Connection
Water Pressure

**Figure 12-2** Wet pipe sprinkler system.

13. a. 2
    b. 1
    c. 4
    d. 3

## True or False

1. a
2. b
3. a
4. b
5. a
6. a
7. a
8. a
9. b
10. a

## Multiple Choice

1. b, d–any order is acceptable
2. d
3. b
4. b
5. b, c–any order is acceptable
6. b, c–any order is acceptable
7. a
8. d
9. c
10. a
11. c
12. b

13. d

14. a

15. d

16. c

17. c

18. b

19. c

20. d

21. c

22. a

23. d

## Fill-in-the-Blanks

1. b

2. d

3. a

4. e

5. c

6. a. upright

   b. pendent

   c. side wall

7. a

8. a

# CHAPTER THIRTEEN ANSWERS

## Matching

1. c

2. d

3. a

4. e

5. b

6. a. compression

   b. tension

   c. shear

7. a. gable

   b. hip

   c. intersecting

   d. gambrel

   e. mansard

   f. butterfly

   g. shed

8. b

## True or False

1. b
2. a
3. a
4. b
5. a
6. a
7. b
8. a
9. a

## Multiple Choice

1. d
2. c
3. a, d–any order is acceptable
4. a
5. b
6. c
7. d
8. c
9. a
10. b
11. c
12. d
13. a
14. a
15. b
16. a

## Fill-in-the-Blanks

1. b
2. e
3. d
4. a
5. c
6. truss
7. gusset plates

# CHAPTER FOURTEEN ANSWERS

## Matching

1. d
2. c
3. e
4. a
5. b
6. a. extend/retract
   b. raise/lower
   c. rotate
7. a. suitcase
   b. shoulder
   c. flat

## True or False

1. a
2. a
3. b
4. a
5. a
6. b
7. a
8. b
9. b
10. a
11. a
12. b
13. a

## Multiple Choice

1. c
2. b
3. d
4. b, d–any order is acceptable
5. a
6. a, d–any order is acceptable
7. c
8. d
9. a
10. a
11. d
12. a

13. b

14. a, c–any order is acceptable

15. c, d–any order is acceptable

16. d

17. b

## Fill-in-the-Blanks

1. b

2. e

3. d

4. a

5. c

6. a. tip

   b. fly section

   c. pulley

   d. halyard

   e. rungs

   f. rails

   g. ladder locks

   h. gusset plates

   i. guides

   j. bed section

   k. beam

   l. butts

7. A-frame hoist

8. hooking tools

# CHAPTER FIFTEEN ANSWERS

## Matching

1. c

2. e

3. a

4. d

5. b

6. a. round turn

   b. bight

   c. loop

7. a. half hitch

   b. double becket bend

   c. bowline

   d. figure eight

## True or False

1. b
2. a
3. b
4. a
5. a
6. b
7. a
8. a
9. b
10. a
11. a
12. a
13. b

## Multiple Choice

1. c
2. a
3. c
4. c
5. c
6. a
7. d
8. a, c–any order is acceptable
9. d
10. c

## Fill-in-the-Blanks

1. c
2. e
3. a
4. b
5. d

6.

Working End

Standing Part

Running End

**Figure 15-3**  Three parts of a rope.

7. dressed and set

# CHAPTER SIXTEEN ANSWERS

## Matching

1. c
2. d
3. a
4. e
5. b
6. a. pancake
   b. V-type
   c. lean-to
7. a. firefighter's carry
   b. extremity carry
   c. seat carry

## True or False

1. b
2. a
3. b
4. b
5. a
6. b

7. a

8. b

9. b

10. a

11. b

12. a

13. b

14. a

15. b

16. b

17. b

## Multiple Choice

1. b

2. a, b–any order is acceptable

3. c

4. b

5. d

6. d

7. b, c–any order is acceptable

8. c

9. d

10. b

11. a, d–any order is acceptable

12. b

13. c

14. b

15. a

16. a

17. c, d–any order is acceptable

18. d

19. a

20. c

21. c

22. d

23. a

24. c

## Fill-in-the-Blanks

1. c

2. e

3. b

4. a

5. d

6. a. Maintain manual stabilization while another rescuer checks pulse, movement, and sensation.

b. Apply cervical collar. Place backboard next to the patient. Grasp patient's shoulder, hip, and leg.

c. On the command of the firefighter maintaining head stabilization, roll the patient onto the patient's side.

d. Place the backboard under the patient.

e. On the command of the firefighter maintaining head stabilization, roll the patient onto the backboard. Using a long-axis drag, gently center the patient on the board.

f. Secure the patient using the board straps. Recheck distal pulses, movement, and sensation.

# CHAPTER SEVENTEEN ANSWERS

## Matching

1. b
2. d
3. e
4. a
5. c
6. a. door
   b. wall
   c. jamb
   d. frame
   e. hinges
   f. knob-locking device

## True or False

1. b
2. a
3. b
4. a
5. a
6. b
7. a
8. b
9. a
10. a
11. a
12. b

## Multiple Choice

1. c
2. d
3. c
4. a
5. a, d–any order is acceptable

6. b, c–any order is acceptable
7. a
8. b
9. d
10. c, d–any order is acceptable
11. d
12. c
13. d

## Fill-in-the-Blanks

1. d
2. b
3. a
4. e
5. c
6. rotary, masonry, wood
7. Cut a large inverted V-shaped hole from top to bottom. Then, pull the slats from each side, creating a very large opening.

# CHAPTER EIGHTEEN ANSWERS

## Matching

1. b
2. d
3. e
4. a
5. c

## True or False

1. b
2. a
3. a
4. a
5. b
6. b
7. a
8. a
9. b
10. a
11. a
12. b
13. b
14. a

15. b
16. b
17. a

## Multiple Choice

1. d
2. c, d–any order is acceptable
3. b
4. b
5. a
6. b
7. c
8. a, c–any order is acceptable
9. a
10. c
11. d
12. b
13. c
14. d
15. c
16. a
17. d
18. c
19. b, d–any order is acceptable
20. a
21. c, d–any order is acceptable
22. b

## Fill-in-the-Blanks

1. d
2. e
3. a
4. c
5. b
6. a. primary hole

   b. examination holes

   c. trench cut

7. A trench cut is made in the roof and then the ceiling is pulled to promote vertical airflow through the trench. Additionally, a handline should be in place below the opening to cut off any horizontal extension.

8. The firefighters are using a tool to take out a window. First, they attach a tool to a rope and lower the tool to the window. Next, they wrap a turn of rope around their wrist and then pull up the tool. The final step is to toss the tool as far out from the building as possible and let it fall back into the window.

# CHAPTER NINETEEN ANSWERS

## Matching

1. c
2. e
3. d
4. a
5. b
6. f
7. b
8. a(2); b(3); c(1)

## True or False

1. b
2. b
3. b
4. a
5. a
6. b
7. b
8. b
9. a
10. a
11. b
12. a
13. a
14. a
15. a
16. b
17. b
18. a
19. a
20. a
21. b

## Multiple Choice

1. a
2. c
3. b
4. b

5. a, d–any order is acceptable
6. b
7. d
8. a
9. c
10. a
11. a, d–any order is acceptable
12. c
13. c
14. c
15. b
16. d
17. b
18. d
19. d
20. c
21. b
22. d
23. a
24. c
25. b, d–any order is acceptable
26. c
27. a
28. d
29. d
30. b
31. b

# Fill-in-the-Blanks

1. e
2. j
3. g
4. i
5. b
6. h
7. a
8. c
9. f
10. d

11.

| **Fireground Factors** | | |
|---|---|---|
| **BUILDING** | | |
| Size | Construction type | Condition |
| Age | Openings | Utilities |
| Concealed spaces | Access | Effect of fire |
| Extent of fire | Interior fuel load | Exterior fuel load |
| **FIRE** | | |
| Size | Location | Direction of travel |
| Time since ignition | Extent | Materials involved |
| Material left to burn | Fire load | Stage of involvement |
| **OCCUPANCY** | | |
| Type | Value | Fire load |
| Status (used/vacant) | Hazards of occupancy | Life hazard |
| Arrangement | Obstructions | |

*Note:* Any order within the categories is acceptable.

**Table 19-1** There can be numerous building, fire, and occupancy factors to consider.

12. weather

topography

fuel

13. Heavy timber fires burn with high intensity and flame length.

14. Firefighters apply water fog to cool a venting propane tank.

15.

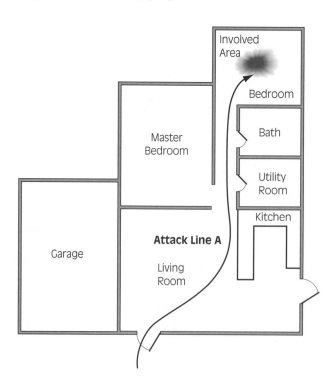

**Figure 19-6** Attack line.

16.

**Figure 19-7** Prioritization of rescues.

17. a. indirect
    b. direct

# CHAPTER TWENTY ANSWERS

## True or False

1. a
2. b
3. b
4. b
5. b
6. b
7. a
8. b
9. a
10. b
11. a
12. a
13. b
14. a
15. b
16. a
17. a

## Multiple Choice

1. c, d–any order is acceptable
2. c, d–any order is acceptable
3. b
4. c
5. a
6. b
7. c
8. d
9. a
10. a
11. b
12. a, d–any order is acceptable
13. a
14. b, d–any order is acceptable
15. d
16. c
17. d
18. a

# CHAPTER TWENTY-ONE ANSWERS

## True or False

1. a
2. b
3. a
4. b
5. a
6. b
7. b
8. a
9. a
10. a
11. b
12. b
13. a
14. a
15. a
16. a
17. a

## Multiple Choice

1. d
2. b
3. b, c–any order is acceptable
4. b
5. d
6. b
7. d
8. a
9. d
10. c
11. d
12. a
13. d
14. c
15. c, d–any order is acceptable
16. b
17. c
18. a, b–any order is acceptable
19. c
20. d
21. b
22. b
23. c
24. c

# CHAPTER TWENTY-TWO ANSWERS

## True or False

1. a
2. a
3. b
4. a
5. b
6. a
7. a
8. b
9. a
10. a

## Multiple Choice

1. d
2. c
3. d
4. a
5. d
6. d
7. a, b–any order is acceptable
8. d
9. b
10. b
11. a
12. d
13. b
14. d
15. a, b–any order is acceptable
16. d

## Fill-in-the-Blanks

1. b
2. e
3. a
4. c
5. d
6. a. carotid
   b. brachial
   c. radial
7. a. dilated
   b. constricted
   c. unequal
8. a. partial thickness
   b. full thickness
   c. superficial

# CHAPTER TWENTY-THREE ANSWERS

## True or False

1. b
2. a
3. a
4. b

5. a

6. a

7. b

8. a

9. b

10. b

11. b

12. a

13. a

14. a

15. a

## Multiple Choice

1. d

2. b

3. b, c–any order is acceptable

4. a

5. b

6. c

7. a, d–any order is acceptable

8. d

9. a

10. b, d–any order is acceptable

11. c, d–any order is acceptable

12. b

13. d

14. a

15. c

16. a, d–any order is acceptable

17. b

# CHAPTER TWENTY-FOUR ANSWERS

## True or False

1. a

2. b

3. a

4. a

5. b

## Multiple Choice

1. d
2. c
3. b
4. a
5. b
6. c
7. c
8. d
9. c
10. b
11. c
12. c
13. b
14. d

## Fill-in-the-Blanks

1. c
2. e
3. b
4. a
5. d

# CHAPTER TWENTY-FIVE ANSWERS

## Matching

1. b
2. e
3. a
4. c
5. d
6. a. pressurized gas
   b. nonpressurized liquid
   c. corrosives

## True or False

1. a
2. a
3. b
4. b
5. b
6. a
7. b

8. a

9. a

10. a

11. a

12. b

13. a

14. b

15. a

## Multiple Choice

1. b

2. c

3. c

4. d

5. b

6. a

7. a

8. d

9. c

10. b

11. a

12. d

13. d

14. c

15. a

16. c

17. d

18. d

19. c

20. d

21. d

22. a

23. d

24. b

25. c

## Fill-in-the-Blanks

1. c

2. a

3. e

4. d

5. b

# CHAPTER TWENTY-SIX ANSWERS

## True or False

1. a
2. b
3. b
4. a
5. b
6. a
7. b

## Multiple Choice

1. c
2. b
3. a, d–any order is acceptable
4. d
5. a
6. d
7. c
8. d
9. c
10. d
11. b
12. c

# CHAPTER TWENTY-SEVEN ANSWERS

## Matching

1. b
2. e
3. d
4. a
5. c
6. a. inhalation
   b. absorption
   c. ingestion
   d. injection
7. b

## True or False

1. b
2. a
3. b
4. b
5. b
6. b
7. a
8. b
9. b
10. a
11. b

## Multiple Choice

1. d
2. c
3. b, d–any order is acceptable
4. b
5. d
6. a
7. c
8. d
9. b
10. d
11. b
12. c
13. a

# CHAPTER TWENTY-EIGHT ANSWERS

## Matching

1. e
2. b
3. h
4. f
5. c
6. g
7. a

8. i
9. d
10. j
11. a

## True or False

1. b
2. a
3. a
4. b
5. a
6. a
7. a
8. b
9. a
10. a
11. a
12. b
13. a
14. b
15. a
16. b

## Multiple Choice

1. a
2. c
3. b
4. d
5. b
6. a
7. d
8. a, b—any order is acceptable
9. c
10. c
11. b, c—any order is acceptable
12. b
13. d
14. c, d—any order is acceptable
15. a
16. b, d—any order is acceptable
17. c
18. d
19. b